MW00754126

3D Printing and CNC Fabrication with SketchUp

About the Author

Lydia Sloan Cline teaches drafting, digital modeling, and 3D printing classes at Johnson County Community College in Overland Park, Kansas. She works for architecture firms, judges competitive technology events, and became a SketchUp fan after using its Push/Pull tool for the first time. Lydia is also the author of *3D Printing with Autodesk 123D*, *Tinkercad*, and *MakerBot* (published by McGraw-Hill Education) and other books.

3D Printing and CNC Fabrication with SketchUp

Lydia Sloan Cline

New York   Chicago   San Francisco   Athens   London   Madrid
Mexico City   Milan   New Delhi   Singapore   Sydney   Toronto

Cataloging-in-Publication Data is on file with the Library of Congress

McGraw-Hill Education books are available at special quantity discounts to use as premiums and sales promotions or for use in corporate training programs. To contact a representative, please visit the Contact Us page at www.mhprofessional.com.

3D Printing and CNC Fabrication with SketchUp

Copyright © 2016 by McGraw-Hill Education. All rights reserved. Printed in the United States of America. Except as permitted under the United States Copyright Act of 1976, no part of this publication may be reproduced or distributed in any form or by any means, or stored in a database or retrieval system, without the prior written permission of the publisher.

McGraw-Hill Education, the McGraw-Hill Education logo, TAB, and related trade dress are trademarks or registered trademarks of McGraw-Hill Education and/or its affiliates in the United States and other countries and may not be used without written permission. All other trademarks are the property of their respective owners. McGraw-Hill Education is not associated with any product or vendor mentioned in this book.

1 2 3 4 5 6 7 8 9 0   DSH DSH   1 2 1 0 9 8 7 6 5

ISBN 978-0-07-184241-9
MHID   0-07-184241-1

This book is printed on acid-free paper.

Sponsoring Editor
  Michael McCabe

Copy Editor
  James Madru

Editorial Supervisor
  Stephen M. Smith

Proofreader
  Claire Splan

Production Supervisor
  Lynn M. Messina

Indexer
  Claire Splan

Acquisitions Coordinator
  Lauren Rogers

Art Director, Cover
  Jeff Weeks

Project Manager
  Patricia Wallenburg, TypeWriting

Composition
  TypeWriting

Information contained in this work has been obtained by McGraw-Hill Education from sources believed to be reliable. However, neither McGraw-Hill Education nor its authors guarantee the accuracy or completeness of any information published herein, and neither McGraw-Hill Education nor its authors shall be responsible for any errors, omissions, or damages arising out of use of this information. This work is published with the understanding that McGraw-Hill Education and its authors are supplying information but are not attempting to render engineering or other professional services. If such services are required, the assistance of an appropriate professional should be sought.

To Makers all over the world and the companies that support their efforts.

And a big thanks to…

Roger, Amber, and Christie, who indulge and contribute to my own Maker efforts.

Contents

Preface

THE PAST DECADE has seen tremendous interest and possibilities in 3D printing and CNC fabrication. Novices and professionals are using them to make their lives and jobs more rewarding and productive. When these technologies are combined with websites such as Quirky, Kickstarter, Prosper, Etsy, and Shapeways, and social media outlets such as Twitter and Facebook, people are empowered to turn hobbies into businesses. 3D printing, CNC fabrication, and those who do them are loudly ringing in the New Industrial Revolution.

3D printing in particular is becoming part of our nation's education, and to be competitive in school and work most people will eventually require knowledge of it. Electronics, toy, food, automotive, and construction companies are finding innovative ways to use this technology. The military is experimenting with it. 3D printing solutions are currently being utilized in everything from housing to human body parts. But even so, current 3D printers are comparable to the dot-matrix printers of the early 1980s. Eventually they'll enable solutions we can barely imagine right now.

Creating, manufacturing, and advertising are becoming democratized because of affordable tools and low startup costs. No matter what your age or background, you can participate, because the barriers are low. The same way that software and laser printers let anyone publish from their kitchen table, and WordPress and a website let anyone have a nice storefront, free modeling software, cheap printers, and community Maker spaces enable anyone to put out a product. Mass manufacturing isn't going away, but micromanufacturing and just-in-time manufacturing that cater to niche and customization markets also will find a bigger place (Figure 1). So will the Makers who learn these technologies.

Websites such as the Trimble 3D Warehouse, where designs are freely shared, let the world's citizens tinker with, customize, and improve each

Truss Cuff (sz M)
$65.00 by SliceLab

N12.bracelet
$49.99 by Continuum

Euro 6-in-1 Chainmail Bracel...
$10.00 by vertigopolka

Dilly Design Interlaced Patter...
$19.00 by resewash

WEB BANGLE
$37.42 by JAXJEWELRY

Porous Cuff (sz S/M)
$120.00 by nervoussystem

Bracelet Constructionist Slee...
$38.07 by mcodr

Bracelet I Medium
$60.00 by hulahy

Figure 1   A page at shapeways.com, where Makers print and sell their creations.

other's work in the spirit of open source. Maker Faires let them show off what they've done. The nascent Maker movement might counter cheap overseas labor and bring manufacturing back to the United States. Teachers, students, homeschoolers, dreamers, entrepreneurs, and *you*, since you're holding this book, are making this happen! You may already have some concrete ideas about what you'd like to do. So congratulations on finding your way here, because now you'll learn how to turn those ideas into a physical product.

Let's get started!

Lydia Sloan Cline

Resources to Check Out

- **Autodesk University 2013 video:** *The New Industrial Revolution,* available at: http://au.autodesk.com/au-online/classes-on -demand/class-catalog/2013/class-detail/3604

- **Information about Maker Faires:** http://www.makerfaire.com

- **Kickstarter, Indiegogo, and Prosper, sites where entrepreneurs can solicit crowdfunding, market research, and feedback from a large participant community:** http://www.kickstarter.com http://www.indiegogo.com http://www.prosper.com

- **Quirky, a large community that gives feedback on the product development and manufacturing process:** http://www.quirky.com

- **Etsy, an online marketplace for handmade products:** www.etsy.com

- **Shapeways, Ponoko, and Imaterialise, online service bureaus where you can send files for 3D printing or CNC cutting and sell copies via their storefronts:** http://www.shapeways.com http://www.ponoko.com http://i.materialise.com/

- **Trimble 3D Warehouse, a website from which to download all kinds of things to tinker with and 3D print:** https://3dwarehouse.sketchup.com/

- **Instructables, a website from which to get instructions to make all kinds of things:** http://www.instructables.com

- Cline, Lydia Sloan. *3D Printing with Autodesk 123D, Tinkercad, and MakerBot,* 2015 (New York: TAB/McGraw-Hill)

- Cline, Lydia Sloan. *SketchUp for Interior Design: 3D Visualizing, Designing, and Space Planning,* 2014 (Hoboken, NJ: Wiley)

CHAPTER 1

We 🖤 Creating!

HELLO! SINCE YOU FOUND your way here, I'm betting you're a creator, someone who sews, cooks, crafts, woodworks, builds, machines, programs, repurposes, fixes, invents, and tinkers. Perhaps you have or want a small business and are always thinking up new wares to make. Maybe you make cosplay items. You might be on a quest for the Next Best Thing or a way to improve an existing thing. Or maybe you just want to decorate your life with tangible outputs of your ideas (Figures 1-1 and 1-2). You probably look at items in your field of interest and consider how to create similar or better ones. You discuss your ideas with the similar-minded and attend gatherings where ideas and wares are shown off (Figure 1-3). There's

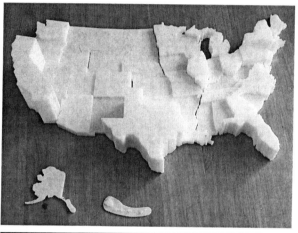

Figure 1-2   A puzzle made with SketchUp. (*Courtesy of Chris Krueger, http:// thenewhobbyist.com/.*)

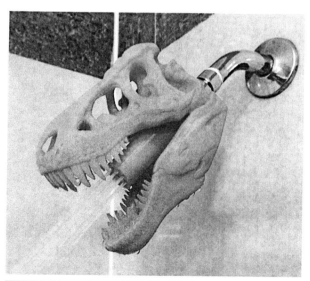

Figure 1-1   A dinosaur showerhead from the Thingiverse.

Figure 1-3   A display at the Kansas City Maker Faire.

actually a word for this—*Maker*. Makers are part of the new industrial revolution, and you've come to the right place!

What Is This Book About?

This book shows how to digitally model your ideas, make those models printable, and then bring them to physical life. It's for those who want to design, iterate, customize, and prototype their ideas—jewelry, shoes, eyeglasses, phone accessories, toys, promotional materials, furniture, whatever—relatively easily and cheaply. To do so, we'll make some fun, easy projects with a popular program called *SketchUp*, pairing it up with other programs as needed. We'll also physically make some of our projects on a 3D printer and on a CNC (computer numerical control) router. We'll discuss how to operate, maintain and troubleshoot a 3D printer, and software that runs printers and routers. If you don't personally own such machines, no worries, we'll also discuss places you can go to get your creations made.

No prior knowledge of any drafting or modeling programs is assumed or needed. My assumption is that you have no modeling or 3D printing experience and are using SketchUp for the first time. Hence, everything is explained in a project-based, step-by-step manner. So if you're a beginner, you've come to the right place! This book is arranged chronologically, and the content builds from the first chapter to the last. Make sure to check out my YouTube channel, where videos of some of this book's projects are posted. Its URL is at the end of this chapter.

My goal is to get you get up and running fast in the 3D modeling, printing, and fabrication world. Hence, our use of SketchUp's tools and functions will be limited to those relevant to 3D printing projects. So know that while SketchUp has the capability to do such things as create construction documents and animations, we won't be doing that.

What Is a Modeling Program?

A *modeling program* is graphics software in which you create three-dimensional (3D) drawings called *models*. You can spin these drawings around on the screen to view from any angle. Modeling is different from traditional computer-aided drafting (CAD) software in that the latter is basically an electronic pencil with which you draw two-dimensional (2D) pictures. With modeling, your picture is always 3D.

Many free, ready-made models are available for download, but knowing how to operate modeling software enables you to create your own designs. Maybe you'd like to experiment with some Fitbit and GoPro accessories or customized Lego blocks. Or you need to tweak some premade content to make it printable. This must be done inside modeling software.

There are many modeling programs on the market; besides SketchUp, popular ones include the Autodesk 123D apps, Inventor, 3dsMax, Maya, Rhinoceros, Solidworks, Blender, Modo, Sculptris, and Zbrush. However, most of those have steep learning curves or high prices. The advantages of SketchUp are that it has a free version and you can start modeling your ideas with it pretty fast.

What Is SketchUp?

SketchUp, formerly owned by Google and now owned by Trimble, is a program used to electronically sketch ideas three-dimensionally, or "get your doodle on." It's the closest you can get to pencil and tracing paper for thinking out ideas. You can "sketch" loosely (meaning without imputing numbers) or sketch precisely.

You also can start out sketching loosely and then scale the model precisely.

Specifically, it's a polygonal surface-modeling program. *Polygonal* means that the model is made of mesh, that is, a bunch of polygons (flat surfaces). *Surface* means that the model is hollow (Figure 1-4). Some programs create solid models, which are forms completely filled inside. Both model types have their strengths and weaknesses. Generally, mesh models are best for organic, free-form subjects, and solid models are best for mechanical pieces. That said, SketchUp in particular was designed originally for architects and interior designers to model houses; hence its native (built-in) tools create rectangular items best. Curved items can be made, but within limits; for instance, you can't model facial features very easily. However, the fact that it's used by people in such diverse fields such as game development, animation, logo design, woodworking, catalog illustration, and landscaping means that you'll probably be able to do what you want with it, too. SketchUp

is a not a parametric program that lets you model the object and change all its individual dimensions later. Nor is it a building information modeling (BIM) program, where you can extract volumetric data from it. That said, third-party developers are constantly coming up with downloadable tools that give it some parametric and BIM capabilities.

SketchUp is a *vector* program, meaning that it creates vector files. A *vector file* is a collection of lines and curves that scale up or down without loss of quality. Examples are .pdfs (Adobe documents) and .dwgs (AutoCAD documents). This is as opposed to a *raster file*, which is made with pixels and loses resolution quality when enlarged. Examples are .jpgs and .gifs. SketchUp is strictly a desktop program, not a Web-based one.

There are two versions: a free one called *Make* and a pay one called *Pro*. Both look and work identically; Pro simply has additional features, some of which we'll discuss in this book. Pro also allows commercial use.

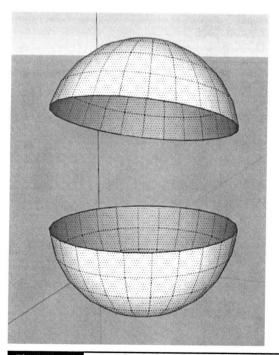

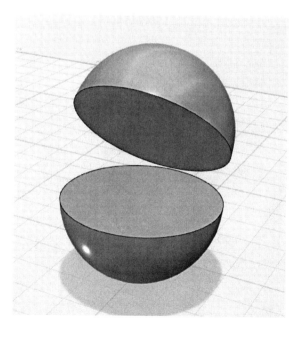

Figure 1-4   SketchUp (*left*) creates hollow polygonal forms. Circles and spheres are actually made of multiple polygons. A solid-modeling program (*right*) creates true forms that are filled inside.

SketchUp works on PC, Mac, and Android platforms and works similarly on all three. Files transfer seamlessly between all platforms. The projects in this book are done on a PC. Mac screenshots are shown in the rare instance when things look or work differently.

SketchUp works well with other programs as part of a larger workflow. We'll use it with Autodesk's AutoCAD (which has a free trial version), 123D Meshmixer (free), and 123D Make (free). You'll learn how to import an AutoCAD file into the Pro version, evaluate a file for printability with 123D Meshmixer, and turn a model into a pattern file for a CNC router with 123D Make. We'll also download and install extensions, which are downloadable tools that let you do things SketchUp's native (built-in) tools don't. Some are free; some have a cost. We'll discuss where to find them, how to install and use them, and any costs and licensing restrictions as we get to them.

machine that builds it. It does this by applying successive layers of material onto a flat surface called a *build plate* and melting that material into the shape of the model. You can assemble a printer yourself or buy one preassembled. Consumer model prices range from around $300 for unassembled parts to around $6,000 for a higher-end assembled machine. 3D printers plug into a USB port on your computer just like any other peripheral (Figure 1-5). They're run with software that either comes with them or that can be downloaded for free or cost from various sites.

Printers use *filament*, a string of material wrapped around a spool (Figure 1-6). Consumer machines primarily use thermoplastic, the main types being PLA (plant-based) and ABS (oil-based). Both come in all colors. Other filament types, such as flexible, dissolvable, conductive, and wood, are also available. Commercial machines use a wide range of filaments, such as

What Is 3D Printing?

3D printing is the process of making a physical model from a digital one. A 3D printer is the

Figure 1-5   A 3D printer plugs into a computer USB port.

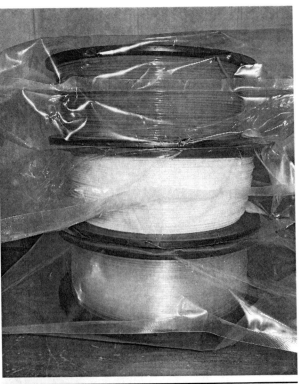

Figure 1-6   Plastic filament spools.

Figure 1-7 A printer making arteries with biomedical filament.

Figure 1-8 3D printing doesn't offer economies of scale in time or material. It takes 12 times as long to print 12 items as it does to print one.

ceramic, food paste, silver, gold, and steel. There are even biomedical filaments for making human body parts (Figure 1-7).

Is 3D Printing Cheap?

It depends on what you mean by "cheap"! If you're iterating or prototyping an idea, it's certainly cheaper than having to commission parts made with an injection-mold process. You can have a copy of your idea in the time it takes to print it. However, there's no economy of scale, time- or material-wise. You cannot print multiple copies in the same time as one (Figure 1-8). If you use a third-party printer, you'll be charged for the amount of material the model requires, the time it takes in the printers, and any

post-production labor involved. When you print something on your own machine, the software that interfaces with it will tell you the printing time and amount of material required.

What Is CNC Fabrication?

CNC fabrication is the process of manufacturing objects on a computer-controlled cutting machine. Such machines are typically routers or lasers, and they carve and cut shapes out of sheets of cardboard, plastic, wood, wax, and foam, and blocks of aluminum (Figure 1-9). This is popular for creating furniture, crafts, puzzles, and toys (Figure 1-10). Examples are the Cricut, which is a personal CNC machine that cuts card stock and paper, and the Carvey, which is a personal CNC machine that cuts wood. ShopBot and Tormach make personal machines, and Haas makes mills, which are large, commercial machines. Figure 1-11 shows a Carvey and a Haas.

Figure 1-9 CNC routers cut flat shapes out of sheets of material. On the left are plastic and paper sheets; on the right are aluminum blocks.

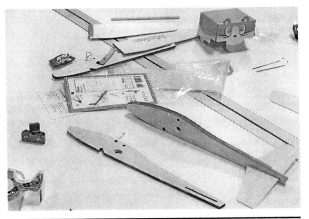

Figure 1-10 Toys are common CNC-milled items.

Figure 1-11 Carvey CNC router (*left*) and Haas CNC mill (*right*).

What Is a Maker Space?

3D printers and CNC machines are too expensive for many Makers, but fortunately, they can be found in *Maker spaces*, also called *hacker spaces*. These are physical locations that contains computers and fabrication tools for member or public use (Figure 1-12). Many libraries and museums have set up these spaces, too. There are TechShops in some states; a monthly membership fee accesses their equipment. Then there's 100Kgarages.com, a matching service that connects people who own shop equipment with people who want to use it. Google "Maker space" or "hacker space" and the name of your city to find what's available near you. There are also service bureaus to which you can upload your files and get them printed for a cost.

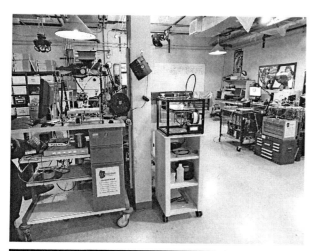

Figure 1-12 HammerSpace, a Maker space in Kansas City, MO.

What You Need, Computer-wise

SketchUp is a graphics-intensive program, so the stronger the video card, the better. Intel cards aren't recommended; NVidia cards work well. You also need a three-button scroll-wheel mouse. A one-button mouse or laptop track pad is usable but difficult. The scroll wheel performs zoom and pan functions, saving you from having to click on those icons all the time. Mac users, your one-button mouse can be swapped with any manufacturer's two-button scroll-wheel mouse. There's also a mouse specifically designed for the 3D environment called a *Space Navigator*. It combines the zoom, pan, and orbit navigation tools, plus it tilts, spins, and rolls (Figure 1-13).

Some Makers use a drawing tablet (not the same as a computer tablet) and digital pen. The tablet's buttons can be programmed to perform multistep functions that otherwise take multiple keystrokes to perform, and the pen offers more control than a mouse. Wacom's Graphire, Intuous, Cintiq, and Bamboo are popular tablets.

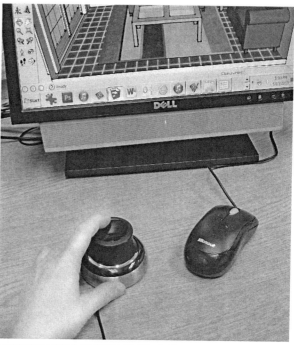

Figure 1-13 The Space Navigator mouse by Logitech is designed for the 3D environment. It can be used along with a traditional mouse.

Here's the recommended hardware for SketchUp 2015:

PC

- Microsoft Internet Explorer browser 9.0 or higher

- SketchUp Pro requires .NET Framework version 4.0.

- 2-GHz or greater processor

- 8 GB or more of RAM

- 500 MB of available hard-disk space

- 3D class video card with 1 GB or more of memory (The video card driver must support OpenGL 2.0 or higher and be up to date.)

- Both the 32- and 64-bit versions of Windows are okay.

Mac OS X 10.10+ (Yosemite), 10.9+ (Mavericks), and 10.8+ (Mountain Lion)

- QuickTime 5.0 for multimedia tutorials

- Safari browser

- 2.1-GHz or greater Intel processor

- 8 GB or more of RAM

- 500 MB of available hard-disk space

- 3D class video card with 1 GB or more of memory (The video card driver must support Open GL version 2.0 or higher and be up to date.)

SketchUp 2015 doesn't work on Windows Vista, Windows XP, OS X 10.7 (Lion), or earlier Mac versions. It doesn't support Boot Camp, VMWare, or Parallels (these are virtual PC programs that run on the Mac). Nor can it be run on a tablet computer. However, viewer programs that enable the model to be seen on iOS and Android tablet computers are available both from SketchUp and from third-party providers. Finally, some of SketchUp's features require an Internet connection. When accessing the Warehouse sites from within SketchUp, only Internet Explorer and Safari work.

Download SketchUp

Before we leave this chapter, let's download SketchUp (Figure 1-14). Point your browser to www.sketchup.com, and go get it!

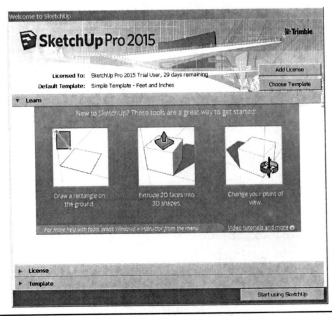

Figure 1-14 The initial screen after the software has been installed.

Windows users, make sure that you choose the appropriate 32- or 64-bit version. Follow the instructions to install. You'll be asked some questions and asked to choose either the free version (Make) or Pro. Even if you choose Make, you'll get a 30-day free trial of Pro. After the end of 30 days, it will revert to Make, but don't worry about your files, they'll still be readable. When you download a future upgrade, it will replace the current one. Different versions can live on your computer, but Make and Pro cannot; one must be uninstalled before the other can be installed. As of this writing, Pro costs $700 for a nonexpiring license. Students who are

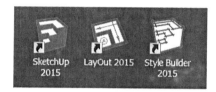

Figure 1-15 When you download SketchUp, two features called *LayOut* and *Style Builder* will install as separate programs with their own icons.

at an accredited school and have a .edu address can obtain a one-year Pro license for $49. This license doesn't permit commercial work. A URL for this is at the end of this chapter.

Two features will install with SketchUp as separate programs: LayOut, which lets you make construction documents, and Style Builder, which lets you customize the digital model's style (Figure 1-15). We won't be using them here, but you may want to investigate them on your own if you choose to go further in SketchUp.

Click on SketchUp's icon to launch. A splash screen will appear, and it will reappear each time you launch SketchUp until either you enter a license number (a Pro purchase) or the trial period ends. At its bottom is a collapsed window called *Template*. Templates are files with default settings, such as measurement units. 3D printers read millimeters (mm), and modeling in millimeters is the most accurate. However, if you're used to using imperial units, choose Simple Template—Feet and Inches (Figure 1-16) because you can convert the model's units

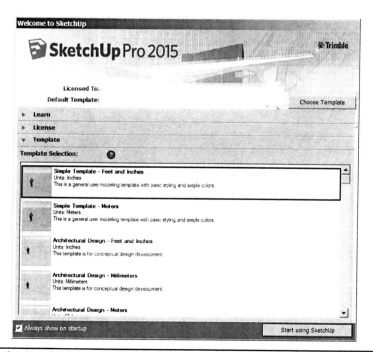

Figure 1-16 Choose the Simple Template—Feet and Inches.

when finished. You can also override the default measurement anytime while you work.

Each time you launch SketchUp, it will open the last template used, but that can be changed, too. SketchUp has a 3D printing template, and we'll use it when we're ready to print something. Making your own templates is possible and discussed in Chapter 2.

Summary

In this chapter we learned what SketchUp is, the computer specs needed to run it, and where to download it. We also learned what 3D printing and CNC fabrication are, what will be covered in this book, and what other programs we'll use as part of a total workflow. Ready to get started? Head over to Chapter 2, where we'll tour the SketchUp interface.

Resources

- **Lydia's YouTube channel:**
  https://www.youtube.com/user/ProfDrafting
- **Official SketchUp blog:**
  http://blog.sketchup.com/

- **Qualified students can obtain a low-cost, one-year license for Pro here:**
  http://www.creationengine.com/html/m.lasso?m=GG&gclid=CLe-nqe-n8cCFQswaQod9scOtQ
- **Space Navigator mouse website:**
  http://www.3dconnexion.com/products/spacenavigator.html
- **SketchUp hardware requirements:**
  http://help.sketchup.com/en/article/36208
- *SketchUp for Interior Design*, by Lydia Sloan Cline (Hoboken, NJ: Wiley, 2014). Despite the title, it's a general SketchUp book with examples geared to architectural spaces and furniture. It discusses many SketchUp features not covered in this book.
- **General Electric's traveling-garage website, where Makers can view and use its prototyping and manufacturing tools:**
  http://www.ge.com/garages
- **TechShop's website:**
  http://www.techshop.ws/
- **100K Garages:**
  http://www.100Kgarages.com/

CHAPTER 2

Getting Started: The Interface

IN CHAPTER 1 WE DOWNLOADED and launched SketchUp. In this chapter we'll look at its interface and how to navigate it.

The SketchUp Interface

The interface consists of a Menu bar, modeling window, three axes, Scale figure, Instructor box, Measurements box, and four circled symbols (Figure 2-1).

Menu Bar

This is a horizontal bar at the top of the screen that houses tools and functions. It has eight categories: File, Edit, View, Camera, Draw, Tools, Window, and Help. An entry called Extensions will also appear after we add our first one.

- **File.** This contains the standard open, save, print, import, and export functions plus some SketchUp-specific ones such as

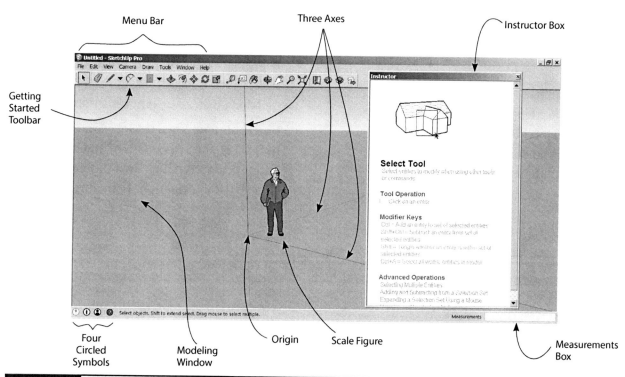

Figure 2-1 The SketchUp interface.

accessing the 3D Warehouse and Google Earth data.

- **Edit.** This contains the standard cut, copy, and paste functions plus SketchUp-specific ones such as paste-in-place, groups, and components.

- **View.** This contains functions that alter how the model looks, such as making it monochrome or wireframe and hiding or displaying features.

- **Camera.** This contains tools that change your position relative to the model and let you view it two- or three-dimensionally.

- **Draw.** This contains tools that make lines, arcs, and shapes, the basis of modeling.

- **Tools.** This contains tools that edit the model, such as erase, move, and scale.

- **Window.** This contains functions that alter the model's properties and dialog boxes that add components and materials and give information about the model.

- **Help.** This contains links to the knowledge center and software updates. Click "Welcome to SketchUp" to access the splash screen, where you can add or change the license number.

The Menu bar on the Mac has one additional entry: a SketchUp menu at the beginning (Figure 2-2). Some of the submenus in the PC's Windows menu are housed here, notably the Preferences submenu.

- **The Getting Started toolbar.** This horizontal bar under the Menu bar contains icons that activate tools. It is one of many toolbars, and we'll activate some more later. Note the drop-down arrows to the right of some tools. They access more tools.

- **The modeling window.** This is the large green and blue (ground and sky) area where you create the model.

- **Three axes.** These are the height, length, and depth lines along which SketchUp draws. Their intersection is called the *origin*. The height (y) line is blue, and the ground lines (x and y) are red and green, respectively. The axes lines are solid in the positive directions and dotted in the negative directions. Every point has a coordinate (group of numbers) that describes its location. (3,4,5) means 3 units along the x/red axis, 4 units along the y/green axis, and 5 units along the z/blue axis. It's best practice to model in the upper-right quadrant and near or on the origin, as shown in Figure 2-3. Number coordinates are positive there, making inputted numbers and calculations easier, plus SketchUp becomes glitchy when you model far from the origin.

- **Human-scale figure.** This helps you to gauge sizes, which is helpful when you're loosely sketching instead of inputting dimensions.

- **Instructor box.** This window contains a brief tutorial of the activated tool and links to online resources. Click on a different tool to change the information. If the Instructor box

Figure 2-2   The PC (*top*) and Mac Menu bars.

distracts you, go to Windows > Instructor, and uncheck it. When you want it back, recheck it.

■ **Measurements box.** This is a field in the lower-right corner of the screen in which all inputted numbers appear. If you don't see it, it's probably hidden behind your Windows taskbar. Maximize the SketchUp screen, or drag it higher on the desktop.

■ **Bottom of form.** Four circled symbols. Click these to make the following pop-up screens appear; reclick to make them disappear:

- **Map pin.** This gives the model's geographic location (if one has been specified).

- **Human figure.** This gives information on the model, such as its properties.

- **Profile figure.** This logs you in and out of your Google account.

- **?.** This turns the Instructor box on and off.

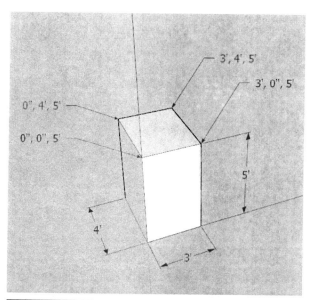

| Figure 2-3 | Every point has a coordinate that describes its location along the three axes. Note the model's location at the origin. |

The Measurements Box

Typing numbers is how to make SketchUp models accurate. You can type anywhere on the screen, but the numbers will appear in the Measurements box. Such numbers include line lengths; circle diameters; rectangle sizes; number of polygon sides; number of copies; distance to move, offset, or push/pull something; and rotation angles.

The nature of the number that appears in the Measurements box depends on which tool is activated. The number could mean inches, degrees, or number of sides. Line lengths are entered like this:

■ **2'** means "2 feet."

■ **2'8** means "2 feet, 8 inches." There is no space between the numbers and symbols.

■ **2'8 1/2** means "2 feet, 8 and ½ inches. There is a space between the 8 and the 1. Alternatively, you could type **2'8.5**.

■ **2 1/2** means "2 and ½ inches." There's a space between the 2 and the 1. Alternatively, you could type **2.5**.

Rectangles require two numbers; typing **48,18** (no space between the characters) makes a rectangle 48 inches long and 18 inches wide. You could also type it **4'18**. The first number goes along the red axis, and the second number goes along the green axis.

Because we selected the Architectural Design—Feet and Inches template in Chapter 1, all numbers will appear in the Measurements box in feet-and-inches format. You can change the template anytime by clicking on Windows > Preferences > Templates (PC) or SketchUp > Preferences > Template (Mac) and selecting a new one. Your current template won't change, but after exiting and reopening the program, the new one will appear.

The nature of the number that appears in the Measurements box depends on which tool is activated. The number could mean inches, degrees, or number of sides. As we cover different tools, how they're entered in the Measurements box will be discussed.

Run Multiple SketchUp Files at the Same Time

On a PC, clicking File > New prompts you to save the current file. On saving, the file closes, and a new instance (open copy) of SketchUp opens. Clicking File > Open also prompts you to save the current file and then navigates to another file. To run multiple instances, right-click on the Desktop icon and select Open or click on the icon of the file you want to open. Know that while you can run multiple instances, you can't run multiple files under one instance.

On a Mac, you can have multiple files open in one instance. Clicking File > Open or File > New opens a new file without closing the current one, which mimics how most other software operates.

Save Options

Under File, there are four save options. Save does just that: it saves the open file. Save As replaces the open file with a new one. Save A Copy As leaves the current file open and makes a closed copy at a location you choose. Save As Template makes a template out of your file. The file is saved with a .skp extension. Files made in earlier versions of SketchUp can be opened in later ones, but files made with later versions cannot be opened in earlier ones. They must be saved as an earlier-version first; do this by scrolling through the Save As Type field at the bottom of the Save dialog box (Figure 2-4).

SketchUp occasionally crashes, so save often. You can program it to automatically save at a time increment of your choice at Window > Preferences > General > Autosave.

PC users exit the software at File > Exit or by clicking the X in the upper-right corner. Mac users go to SketchUp > Quit SketchUp to exit the software; clicking the red button in the upper-left corner just closes the active file, not the software (recall that you can have multiple files open in one instance).

We haven't saved this file yet, so do it now. Call it *Practice*, and choose where on your computer to save it. Perhaps make a folder called *SketchUp Stuff*, and keep all work from this book in it.

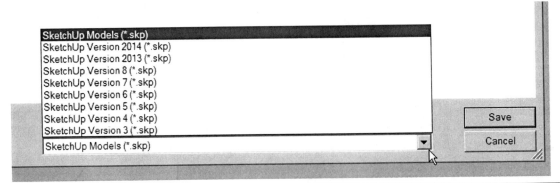

Figure 2-4   SketchUp files can be saved in their current version or an older version using the Save As option.

Backup Files

Backup files are automatically made in the same location as the .skp file (ensure that the Create Backup box is checked at Preferences > General). They have an .skb extension. Don't delete backups until done with a project because they're useful if the .skp file gets corrupted or overwritten. SketchUp's backup files are unique in that they're the next-to-last saved version of the model, not the last saved version. This is handy if you mess up a model after a save and need to backtrack a bit. Convert the .skb file into an .skp file by overtyping the *b* into a *p*.

SketchUp also creates autosave files at specific time increments after the last save (again, you can set these). That file deletes once you save again or exit. If SketchUp crashes, the autosave file remains, giving you almost up-to-date work.

Add the Large Tool Set

The Getting Started toolbar doesn't contain every tool. On a PC, click View > Toolbars; on a Mac, click View > Tool Palettes > Large Tool Set. This opens the Tools window, from which you can activate more toolbars (Figure 2-5). I typically open the Large Toolset, Standard Toolset (it has operating system functions such

as undo/redo), Warehouse (it offers quick access to a Web database of models), Views (it lets you see the model in 2D), and Sections (it lets you cut slices through the model). Run the mouse over a tool's icon to make a tooltip appear that describes what it is (Figure 2-6).

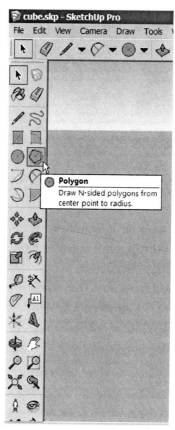

Figure 2-6 The Large Tool Set and the Polygon tooltip.

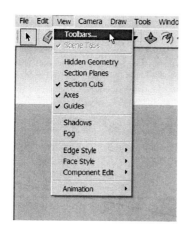
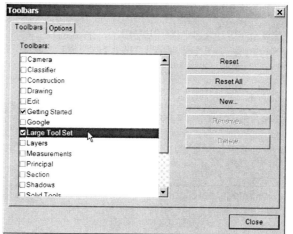

Figure 2-5 At View > Toolbars, check the toolbars you want to add to the workspace.

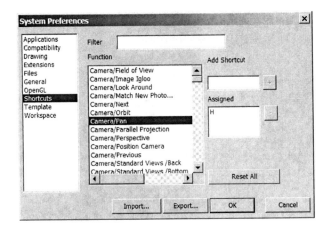

Figure 2-7   Tools can be activated via hotkeys (keyboard shortcuts).

The Large Tool Set duplicates most of Getting Started's tools and has a few others. We'll work with both toolbars open and will access tools through them throughout this book. But they also can be accessed through the Camera, Draw, and Tools menus at the top of the screen if you so choose. Hotkey fans can find a list at Window > Preferences > Shortcuts (Figure 2-7).

Explore! Click on the Menu items, and read the submenus. Run the mouse over each tool icon to read the tooltip. Now let's play with some of those tools.

The Select Tool

The Select tool (Figure 2-8) looks like an arrow. It highlights objects for editing. Click on it to activate.

With the Select tool activated, click on the Human-scale figure. A blue box appears,

meaning that the figure is highlighted and ready for editing. Right-click inside that box and choose Erase from the Context menu that appears (Figure 2-9). He's gone! No worries; go to Edit > Undo Erase and bring him back. Undo reverses the last action, and you can undo all actions one at a time. Then erase him again. You can always reimport a scale figure when needed.

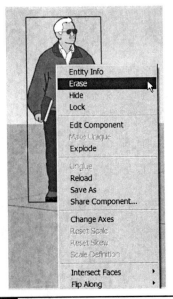

Figure 2-8   The Select tool highlights objects for editing.

Figure 2-9   Right-click on a selected item to bring up a Context menu.

Rectangle and Rotated Rectangle Tools

The Rectangle tool is inside the Shapes menu on the Getting Started toolbar and near the top of the Large Tool Set (Figure 2-10). It makes squares and rectangles that are parallel to the axes. Click it. Then click the cursor anywhere on the Modeling window, drag and release. You've just drawn a rectangle (Figure 2-11).

The Rotated Rectangle tool is below the Rectangle tool in the Shapes menu and next to the Rectangle tool on the Large Tool Set (Figure 2-12). It makes squares and rectangles at any

angle on the workspace; they can be parallel to the axes but don't have to be. It takes three clicks. Click once to place the first corner (a protractor will appear), click a second time to set the angle of the second corner, and click a third time to click the third corner (Figure 2-13). Click the lines on the protractor to set the rectangle sides at 15° increments. Alternatively, type the length and degrees wanted.

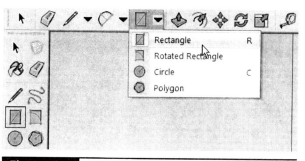

Figure 2-10 The Rectangle tool.

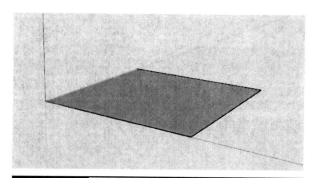

Figure 2-11 A rectangle drawn with the Rectangle tool.

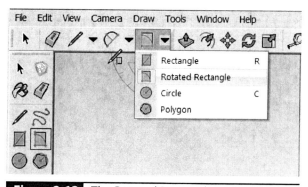

Figure 2-12 The Rotated Rectangle tool.

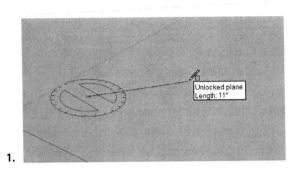

1.

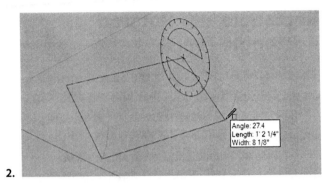

2.

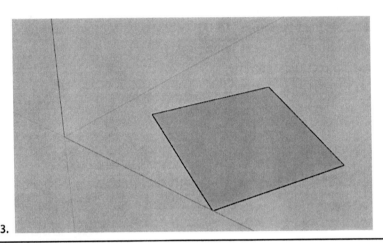

3.

Figure 2-13 Drawing a rotated rectangle.

Faces, Edges, and Geometry

A rectangle is an example of a face. A *face* is a flat plane made of at least three edges (lines). All its edges must be in the same plane ("co-planar") like a piece of paper. A face has two sides: the front, also called the *normal*, and back. The front is white, and the back is a grayish-blue. Recognizing which is front and back is important with 3D printing because the front must face out. Collectively, faces and edges are called *geometry*.

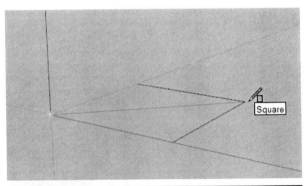

Figure 2-14 A diagonal inference line indicates that the rectangle is currently a square.

The Inference Engine

While dragging the rectangle, you may have noticed that a diagonal dotted line appeared and disappeared (Figure 2-14). This is SketchUp's

inference engine at work. The diagonal line appeared when the rectangle's proportions were dragged into a square shape.

The inference engine is a geometric analysis feature that enables you to draw accurately. Based on how and where you move the cursor, it assumes, or infers, the specific points, planes, and directions you want. You hover over the approximate location of the midpoint, endpoint, edge or center, and other locations, and then the actual midpoint, endpoint, edge, or center appear as a colored dot or line.

> **Tip:** Sometimes the inference you want won't pop up immediately. In that case, encourage it by moving the cursor a few seconds over that area.

The inference engine enables you to make accurate models without having to constantly input dimensions. There are three kinds of inferences: point (e.g., to an endpoint or midpoint), linear (e.g., along the three axes), and planar (e.g., on the model's faces). To clarify planar inferencing, SketchUp "snaps" to a plane when it cannot snap to anything else you've drawn.

As an example of linear dimensioning, see the rotated rectangle in Figure 2-15. In the first graphic, the sides are red and green. This is the inference engine telling you that the sides are parallel to the red and green axes. Modeling parallel to the axes is critical in SketchUp, so watch for these inferences. If a line is black while under construction, it means that it isn't parallel to any axis. While at times this may be

your intent, such as when you draw a rotated rectangle, usually you want your lines parallel to the axes because it's the easiest way to create faces. The biggest problems in SketchUp come from not modeling parallel to the axes. Fixing the resulting defects is tedious, time-consuming, and sometimes impossible. Making a new model is often faster than fixing a defective one.

We'll discuss inferencing more as we use it. For now, just go to Edit > Undo Rectangle (Figure 2-16), and then redraw the rectangle as a square by watching for the diagonal inference line and clicking when it appears.

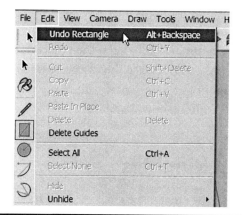

Figure 2-16 Every action can be immediately reversed with the Undo function in the Edit menu.

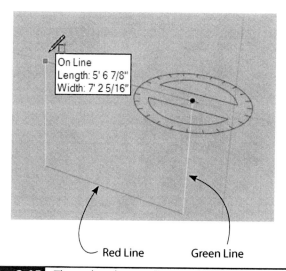

Red Line   Green Line

Figure 2-15 The red and green lines indicate that the sides are parallel to the red and green axes.

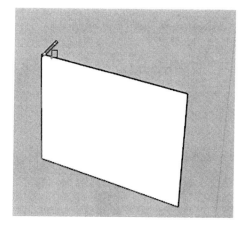

The Push/Pull Tool

The Push/Pull tool (Figure 2-17) adds volume to or subtracts volume from faces by extruding (stretching) them. It's also an autoselect tool, meaning that when you move it onto a face, the face highlights. Not all tools cause faces to highlight; generally, you have to highlight them in a separate action with the Select tool. Click the Push/Pull icon, and then move the mouse onto the square you just drew. See how the face becomes dotted? This means that it's selected and ready to edit. Click, hold, drag the cursor up, and let go. The square is now a cube (Figure 2-18).

Be aware that when the face being extruded is adjacent to another face, it will pull that face along (Figure 2-19). To keep the adjacent face intact, press and release the CTRL key (COMMAND on a Mac) right before performing the push/pull action. A plus sign will appear.

Input Numbers

To make the cube a specific size, type its dimensions as you model it. For example, after making the rectangle, type **5',5'** (no space between the characters) to make it 5 feet on each side. The first dimension goes along the red axis; the second dimension goes along the green axis. Typing must be done immediately after releasing the cursor; if you perform any action in between, it won't work. Right after pushing/pulling, type **5'** to make it that high. You can adjust the cube's dimensions later if needed, but it will take more steps. Because SketchUp's default is inches, the foot symbol must be included or

Figure 2-17   The Push/Pull tool.

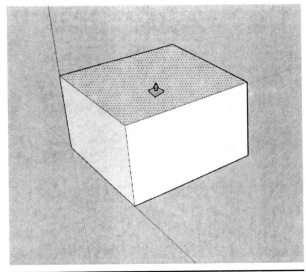

Figure 2-18   A cube made with Push/Pull.

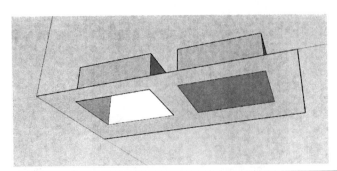

Figure 2-19   To avoid extruding an adjacent face, press and release the CTRL (COMMAND on a Mac) key right before pushing/pulling. Look for a plus sign. On the left is a regular extrusion; on the right is an extrusion when the CTRL key is pressed.

you'll get a 5-inch-tall cube. SketchUp draws at a 1:1 scale, so the 5' is true size. Remember that while 5' appears in the Measurements box, you don't have to type inside that box. You can type anywhere on the screen.

Changing Units

Even though we're using a feet/inches template, you can temporarily input other units by typing the appropriate abbreviation after the number, such as *3m* for 3 meters or *3mm* for 3 millimeters. Imperial and metric units can be mixed and matched in one model. To permanently change the units, go to Window > Model Info > Units (Figure 2-20) and select a different format from the drop-down box.

The Pan, Orbit, and Zoom Tools

The Pan, Orbit, and Zoom tools let you navigate around the workspace. Pan looks like a hand; click on it (Figure 2-21). Then click on the cube, hold, and drag it around the screen. You're panning, that is, moving it around the desktop.

Figure 2-21 The Pan tool.

Figure 2-22 The Orbit tool.

Now click on Orbit, the circular arrows left of Pan (Figure 2-22). This whirls you, the viewer, around the cube; the cube itself doesn't move. Play with this tool—orbit on top, below, and behind the cube (Figure 2-23). If you hold the SHIFT key down while orbiting, you'll temporarily pan. The most efficient way to orbit is to hold the mouse scroll wheel down and drag it around.

Pan is useful to move geometry away from anything overlapping it, such as other geometry or dialog boxes, which are pop-up windows that appear with certain functions. Orbit is useful for spinning a model around to view it from all angles.

The Zoom icon is a magnifying glass to Pan's right (Figure 2-24). Zooming in hones in on the model, letting you see small details. Zooming

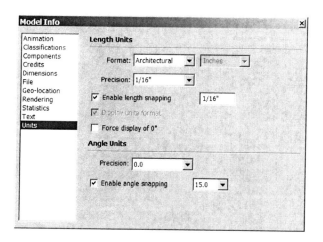

Figure 2-20 To change the model's units, go to Window > Model Info > Units.

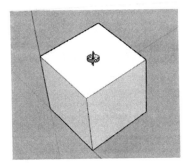
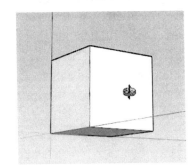

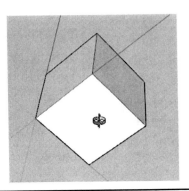

Figure 2-23   Orbiting around the cube.

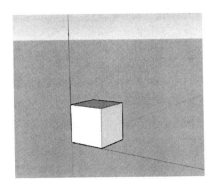

Figure 2-24   The Zoom tool.

out creates a wide-angle view, letting you see the big picture.

Click the Zoom tool onto the model, hold the cursor down, and drag it up and down to zoom in and out (Figure 2-25). Even better, just rotate the scroll wheel on the mouse. The icon to Zoom's right—the magnifying glass with three arrows—is Zoom Extents. Clicking it makes the whole model fill up the window.

If you click Zoom Extents and your model hides off in a corner, it's because you've got little pieces of geometry you drew earlier still lurking around. Find and erase them, and your model will come back. In fact, clicking Zoom Extents is a good way to locate small pieces that you've lost.

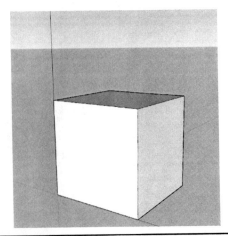

Figure 2-25   Zooming out (*top*) and in (*bottom*).

If your scroll wheel doesn't orbit or pan, the mouse settings may need adjustment. PC users, go to Control Panel > Hardware and Sound > Devices and choose Printer > Mouse. Mac users, go to System Preferences > Keyboard and choose Mouse. Look for a *middle click* wheel setting. If this option isn't there, you may need to install a more recent mouse driver.

Modifier Keys

Modifier keys are keys pressed while using a tool to make it do something else. Examples so far are the SHIFT key while orbiting to temporarily pan and the CTRL key while pushing/pulling to keep a face intact. Here are PC and Mac modifier key equivalents:

PC	Mac
ALT	COMMAND
CTRL	OPTION
ENTER	RETURN
SHIFT	SHIFT
ESCAPE	ESCAPE

The Escape Key

The Escape (ESC) key quits an operation. It cancels dialog boxes and button-less splash screens, closes menus, and quits functions. If you're in the middle of something and need to get out, just hit ESC.

View the Model in Paraline Mode

You've probably noticed by now that the cube you made appears in perspective. That is, it looks the way the eye sees it: small when you're far from it, large when you're close to it, with parallel lines that converge to vanishing points. However, you can display the model as a *paraline* view, which is one where parallel lines don't converge. This is useful when lining features up because you can see them without the distortion of perspective. Click on Camera > Parallel Projection (Figure 2-26). Now you see the cube as an isometric view, a type of paraline drawing where the horizontal lines slope at a 30° angle.

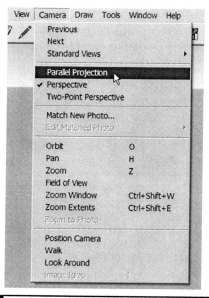

Figure 2-26 Click on Camera > Parallel Projection to see the model in paraline mode. The cube appears as an isometric view, meaning that all parallel lines slope at a 30° angle.

View the Model Two-Dimensionally

You can also make the cube appear two-dimensionally, that is, as top, front, and side views (these are also called *orthographic* views). Keep the parallel projection mode on. On a PC,

Figure 2-27 The Views toolbar.

click on View > Toolbars, and check the Views box. A new toolbar appears with icons that look like 2D views of a house (Figure 2-27). On a Mac, go to View > Customize Toolbar. A large page appears on which tools and toolbars are stored (Figure 2-28). Find and drag the Views toolbar from this window into the Getting Started toolbar.

Click on Views toolbar icons to generate top, front, right, back, and left views (Figure 2-29).

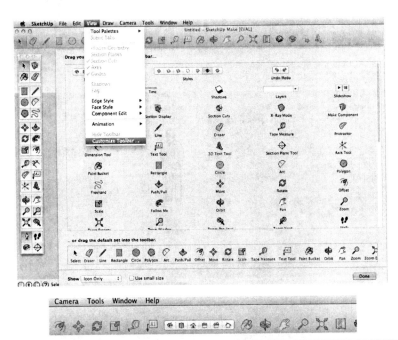

Figure 2-28 On a Mac, clicking on Views > Customize Toolbar brings up a large page on which tools and toolbars are stored (top). Drag the Views toolbar into the Getting Started toolbar (bottom).

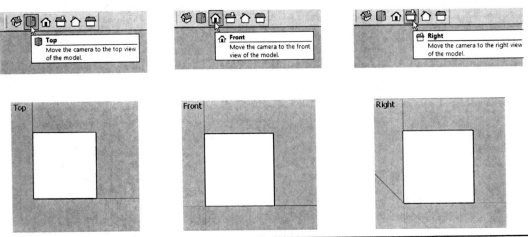

Figure 2-29 Top, front, and left-side views generated from the model.

To return to the isometric view, click the Views toolbar's first icon, the one whose tooltip says "iso." Finally, click Camera > Perspective to return the model to a perspective view. When modeling, align the front of your model (the longest side) with the red axis to make it work best with the Views toolbar.

Selection and Crossing Windows

Geometry must be selected before anything can be done with it. One way to select it is by dragging a window around it (Figure 2-30). Click on the Select tool, and then click on the screen. Hold the mouse down, and drag it from the upper-left corner to the lower-right corner. Let go. This creates a selection window. All geometry entirely within this window will get selected for editing. Anything partly outside

the window will not be selected. Now drag the mouse from the lower-right corner to the upper-left corner. This creates a crossing window. All geometry touched by this window, whether entirely inside it or not, will get selected. The cube is now highlighted, as evidenced by blue dots on its faces, and ready for editing.

A weak video card may cause problems as a model is developed. A common problem is the inability to make selection windows. A quick fix is to go to Preferences > OpenGL and check Use Hardware Acceleration. What you're doing is bypassing the video card and making SketchUp do the calculations itself. While this may work, know that it will also slow the program down when working on large models.

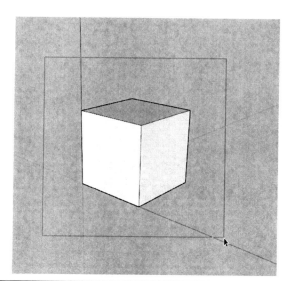
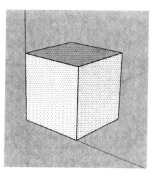

Figure 2-30 A selection window is dragged from the upper-left corner to the lower-right corner and highlights everything inside it. A crossing window is dragged from the lower right-hand corner to the upper left-hand corner and highlights everything it touches. Blue dots indicate a selected face.

Select with Single, Double, and Triple Clicks and with the Shift and Control Keys

Click the Select tool once on a face to select it. Double-clicking selects a face and all its bordering edges. Triple-clicking an edge or a face selects all attached geometry (Figure 2-31). Holding the SHIFT key down brings up a plus or minus (±) sign, indicating that you can add or remove individual pieces from the selection. Holding the CTRL key down brings up a plus (+) sign, indicating that you can add individual pieces to the selection.

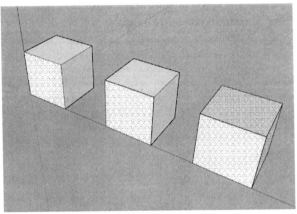

Figure 2-31 Single-click to select a face. Double-click a face to select it and all its edges. Triple-click an edge or face to select everything attached to it.

The Move Tool

The Move tool (Figure 2-32) relocates the model or parts of it. Select the geometry you want to move. Click on the Move tool. Then click on the cube, and move it around! When you move parallel to the axes, lines that color coordinate with those axes appear. As noted earlier, they're inference lines, telling you that you are indeed

Figure 2-32 The Move tool.

parallel to the axis. Figure 2-33 shows the cube moving along the red axis.

You can lock movement along an axis by holding the SHIFT key down after you've moved along the axis wanted (the inference line will become bolded). In this way, you'll be sure to remain parallel to the axis along which you're moving. Any inference can be locked, be it along an axis, along an edge, on a face, from a point, or parallel/perpendicular to an edge.

If whatever you're moving isn't moving parallel to an axes, you can make it do so by pressing an arrow key. This works while using the Move, Line, and Tape Measure tools.

- RIGHT ARROW key = red axis.
- LEFT ARROW key = green axis.
- UP and DOWN ARROW keys = blue axis.

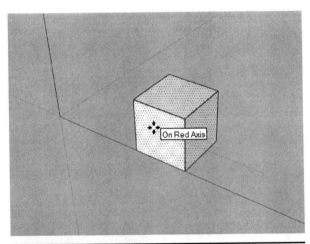

Figure 2-33 A red inference line appears when the cube moves parallel to the red axis.

Copy and Paste

Hightlight the cube, activate the Move tool, and press CTRL. This makes a copy. Alternatively, highlight the entire cube with a selection window or by triple-clicking it. Then press CTRL + C and CTRL + V (on a Mac, press CMD + C and CMD + V). A copy will appear that you can click into place. Click it right next to the first cube

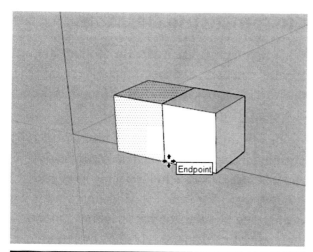

Figure 2-34 Use CTRL + C and CTRL + V to copy highlighted geometry.

(Figure 2-34). You can copy items this way between SketchUp files, too.

Stickiness

Now highlight the copy by double-clicking on one face, holding the SHIFT key down, and then orbiting around the cube and double-clicking on each face (remember that double-clicking will select the face plus its edges). When the whole cube is highlighted, activate the Move tool, click it on the cube, and try to move it. See what happens? It sticks to the first cube and warps it (Figure 2-35). This stickiness of adjacent

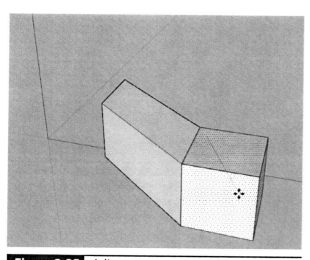

Figure 2-35 Adjacent geometry sticks together.

geometry is the basis on which SketchUp works, and the way you manage it is with groups. Click Edit > Undo until the model returns to one cube.

Groups

Select the whole cube, and then right-click on it. From the Context menu, click on Make Group (Figure 2-36). This puts a bounding box (invisible shell) around the cube. Groups don't stick to anything. They also let you move all the pieces inside them as a whole. Copy the group, and click the copy into place next to the first group. Then move the copy. You'll see that it moves independently without taking the first group with it.

> *Tip:* Make sure when right-clicking any geometry for a Context menu that a tool is active, not Orbit, Pan, or Zoom because they have different Context menus.

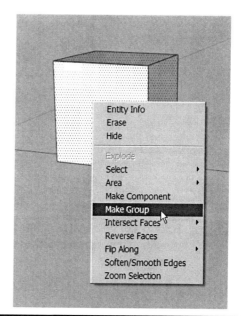

Figure 2-36 Highlight the cube, right-click it, and choose Make Group.

The bounding box should be the same size as the geometry inside it. If it's much bigger, you inadvertently included something else in it. To fix it, right-click and choose Explode from the Context menu, which returns the group to its individual edges and faces. Then carefully select and group it again.

A group must be edited inside its editing box. Double-click on the group to open it. Everything outside it will turn gray (Figure 2-37) and can't be selected. When you're finished editing, click anywhere on the workspace to close the group.

Loose geometry and groups can occupy the same space. Due to this, beginners often forget to open a group's editing box before editing it, resulting in edits being applied outside the group. This becomes evident when the group is moved, and the edits get left behind.

Customize the Desktop

What if you don't like where the toolbars are? What if you don't like their shape? Personalize the workspace by customizing the toolbars and changing their locations.

On a PC, undock and move a toolbar by grasping its handles, the double lines at each end. Change its shape by stretching its edges. Remove a toolbar from the Modeling window by unchecking it in the Toolbars window. You can even move the Measurements box by checking it in the Toolbar window and then dragging it where desired. Toolbars can be moved off the workspace and onto the desktop to free up more modeling room (Figure 2-38).

Figure 2-39 shows how to make a custom toolbar on a PC. Open the Toolbars window, and click New. In the pop-up box, type the name of the toolbar, and hit ENTER. The new toolbar is now listed with all the others. Drag and drop

Figure 2-37 Edits to a group must be made inside its editing box, which is activated by double-clicking.

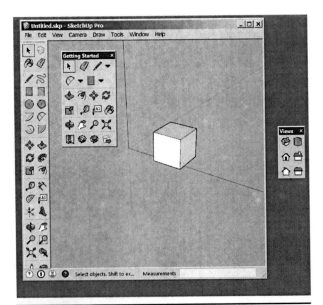

Figure 2-38 Dock and undock toolbars on a PC by grasping their handles and moving them. Stretch to resize. They also can be moved off the workspace.

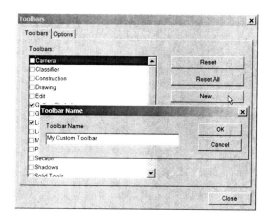

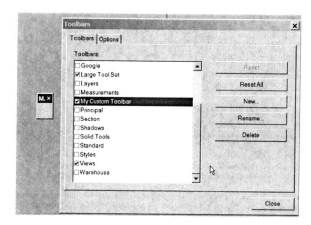

Figure 2-39 Make a custom toolbar by clicking View > Toolbars and choosing New.

tools from other toolbars into it (the Toolbars window needs to remain open while doing this). This enables you to have only the tools you use taking up space on your screen. Drag tools in any toolbar left and right to reposition them (the Tools dialog box must be open while doing this), and click Done to set.

Once you close SketchUp with the toolbars in their new docked positions and shapes, it will remember them each time it opens, along with any screen-size changes. Restore the native toolbars to their original state by clicking Reset on the Toolbars window. If changes don't "take," go to Window > Preferences > Workspace and click Reset (Figure 2-40) after making the desired changes.

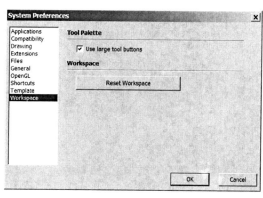

Figure 2-40 At Window > Preferences > Workspace, click Reset Workspace to restore the native toolbars.

Toolbars on a Mac cannot be docked or stretched into different sizes. Customization consists of adding, deleting, and relocating tools on the Getting Started toolbar. Go to View > Customize Toolbar or right-click on the Getting Started toolbar (Figure 2-41). The Tools page we saw earlier appears; drag what you want from it into the Getting Started toolbar. Delete icons from the Getting Started toolbar by dragging them out (they can always be retrieved from the Tools page again). Drag tools left and right to reposition (the Tools dialog box must be open while doing this), and click Done to set.

To restore the Mac Getting Started toolbar to its original state, remove all its tools and then drag the Default Set toolbar at the bottom of the Tools page into it. SketchUp will remember any other Modeling window changes made, such as screen

size and location of other toolbars. To change that memory, make the desired changes, and then go to SketchUp > Preferences > Workspace, and click Reset Workspace (Figure 2-42).

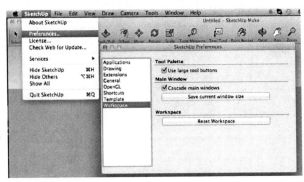

Figure 2-42 To change the memory of the Workspace settings, make the desired changes, go to SketchUp > Preferences > Workspace, and click Reset Workspace.

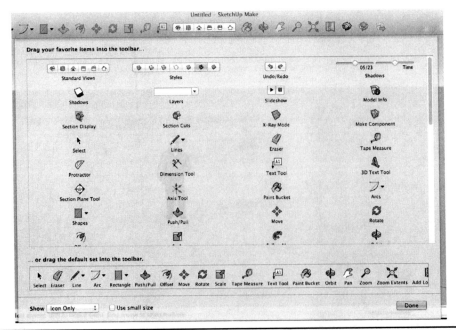

Figure 2-41 Mac users customize the Getting Started toolbar by dragging tools out of it or into it.

Make a Custom Template

A *template* is a file of default settings. Eventually, you may want to create your own. Select File > New to open a new SketchUp file.

- Add and alter toolbar sizes and positions.

- Go to Window > Model Info and adjust whatever settings there you want, such as dimensions.

- Go to Window > Preferences, and adjust whatever settings there you want.

- Draw anything you want all your models to have, such as a customized scale figure, or erase the scale figure if you don't want it.

- Go to File > Save As Template to save this file (Figure 2-43).

Enter a name for this new template file, and save. A thumbnail graphic will appear with the built-in ones at Window > Preferences > Template (Mac users, remember that for you, the

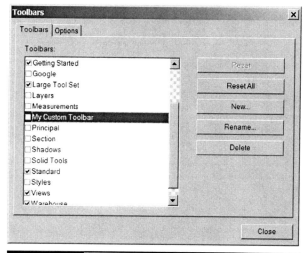

Figure 2-44 Highlight and click DELETE to remove a toolbar.

path is SketchUp > Preferences). Check the Set As Default Template box. This will make it load each time you open SketchUp. To remove it, just highlight and click DELETE (Figure 2-44).

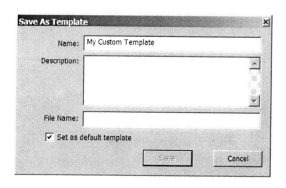

Figure 2-43 Setting up a custom template.

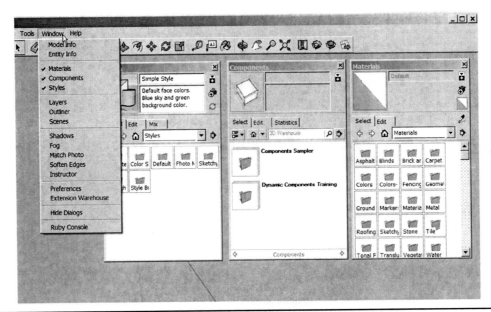

Figure 2-45   The Materials, Components and Styles dialog boxes are shown open.

Manage Dialog Boxes

Earlier we looked at the Instructor box. It's just one of many dialog boxes; open more under the Window menu (Figure 2-45).

As you use them, you'll find that management becomes an issue. Multiple open dialog boxes take up space, obscure the model, and often need to be moved out of the way. Click on the top bar of a dialog box to collapse it; click it again to expand it. Stack multiple boxes together (Figure 2-46), and then click and drag the top box's title bar to move them as a group. Drag a box to remove it from the stack and put it back by snapping it to the bottom of the stack. Keep a collapsed stack of frequently used boxes off to the side, maybe on the computer desktop. SketchUp remembers collapsed dialog boxes, so when you re-open the program, those boxes will probably appear collapsed. Just click them to open.

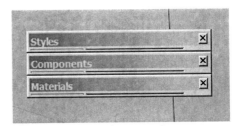

Figure 2-46   Click the top bar of a box to collapse it; click again to expand it. Stack multiple collapsed boxes together to take up less room and move as a group.

Finding Help

Stuck? Click on Help > Knowledge Center. The Help function is online; there is no local help. There's a search box there, but you may have better luck just typing your question directly into Google. Your question probably has already been asked and answered multiple times on one of the many online forums that discuss SketchUp, especially at http://sketchucation.com.

Summary

In this chapter, we toured and navigated the interface and used a few drawing and editing tools. We selected geometry, entered measurements, discussed the inference engine, generated 2D views, and saved the file. We also saw how to personalize the workspace with modified toolbars and custom templates. Now it's time to start modeling! Join me in Chapter 3, and we'll do just that.

Resources

- **Reference card of SketchUp icons and shortcuts:**
  http://help.sketchup.com/en/article/116693

- **SketchUCation, a forum for newbies and experienced users:**
  http://sketchucation.com/

Projects Using SketchUp Make's Native Tools

IN THIS CHAPTER, WE'LL LEARN SketchUp's native (built-in) tools and techniques through seven projects: the icon, a nameplate, chain-link bracelet, pet collar tag, coffee mug, pencil holder, and travel mug. All are modeled with Make, the free version.

Before we start, be aware that all models are not 3D-printable. Just because you can create something on a screen doesn't mean that you can physically print it. For example, a printable model must

- Have a minimum thickness

- Have no sharp edges

- Have no manifold edges (these are edges shared by more than two planes)

- Have no holes, even tiny ones

- Be one solid piece

This chapter, and the next one, mostly focus on learning SketchUp, a task that is enough for now. 3D printability is addressed in detail in Chapter 5.

Before starting each project, go to Views > Toolbars and add the Large Tool Set, Standard, Views, and Warehouse toolbars. Delete the scale figure, and we're ready to start!

Icon Project

In this project, we'll use the Push/Pull, Tape Measure, and Pencil tools and the Divide function to model the SketchUp icon (Figure 3-1).

1. Make a 6' × 6' × 6'7" box (Figure 3-2). Activate the Rectangle tool, and click it onto the workspace. Before clicking a second time, type **6',6'** (no space between the characters). Then press ENTER. Activate Push/Pull, click it onto the rectangle, and type **6'7**. Press ENTER.

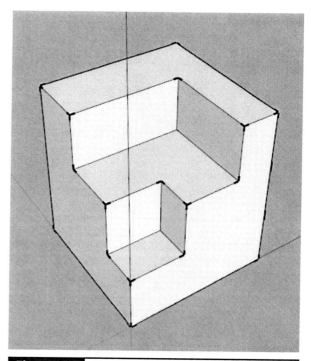

Figure 3-1 The SketchUp icon.

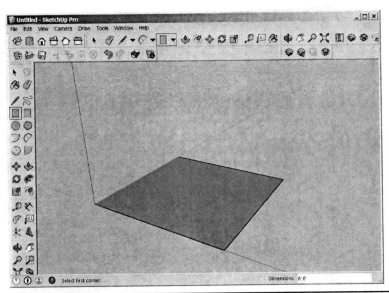
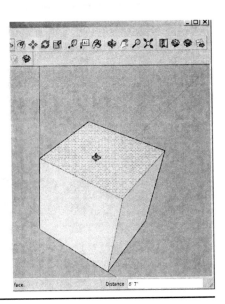

Figure 3-2   Make a 6' × 6' × 6'7" box.

2. Draw a guideline. Activate the Tape Measure tool (Figure 3-3). Click it onto a back edge of the cube's top, and drag it forward. Then type **2'3**, and press ENTER. A guideline will snap to that distance.

Figure 3-3   The Tape Measure tool.

3. Drag a horizontal edge with the Tape Measure to make a guideline (Figure 3-4).

4. Create three more guidelines, all 2'3" apart, for a total of four, as shown in Figure 3-5.

   Trace over the guidelines. Activate the Pencil, also called the Line tool (Figure 3-6), and trace the guidelines shown in Figure 3-7. The Pencil "rubber-bands," meaning that the endpoint of one line is the start of another. If you don't like that

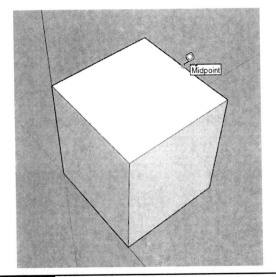
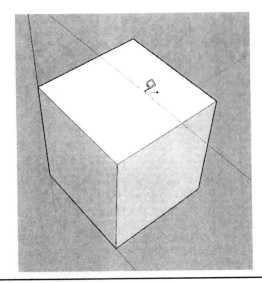

Figure 3-4   Drag an edge with the Tape Measure tool.

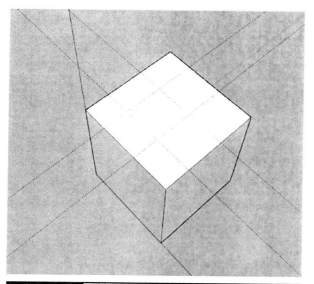

Figure 3-5 Drag four guidelines total, 2'3" apart.

Figure 3-6 The Pencil tool.

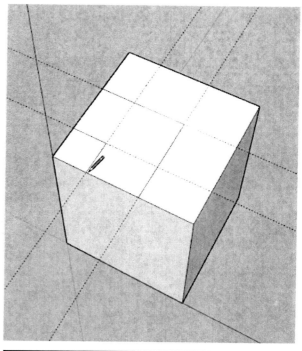

Figure 3-7 Trace over the guidelines with the Pencil.

behavior, go to Window > Preferences, and uncheck Continue Line Drawing. The ESC key also will get you out of the Pencil tool. Always draw parallel to the axes unless the item you're modeling is skewed to the axes deliberately. Holding the SHIFT key down locks the Pencil along an axis.

5. Divide a vertical line into three parts (Figure 3-8). We'll need their endpoints to inference heights. Right-click a line and choose Divide. A bunch of red points along that line will appear. Move the cursor up and down the line until the pop-up window says that

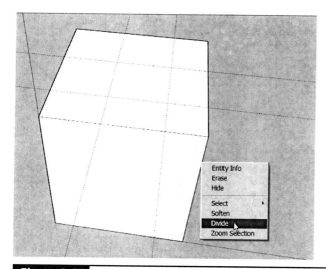

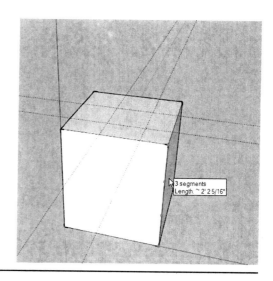

Figure 3-8 Divide a vertical line into three segments.

there arc three segments. Then click on the line to set them.

6. Make the endpoints of the line segments visible (Figure 3-9). Endpoints are turned off by default, so we'll turn them on. Click on Window > Styles and then on the Edit tab. Check the box in front of Endpoints, keeping the default size number at 9.

Line endpoints will appear on the model, including on the endpoints of the three divisions we just made.

7. Push/pull the top planes down, hovering the cursor over the division endpoints (Figure 3-10). The parts will snap to them. This is called *inference matching*.

8. Delete the guidelines by clicking Edit > Delete Guides (Figure 3-11). Done!

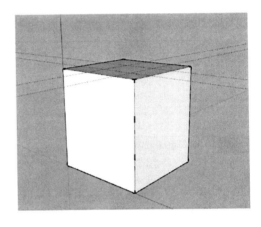

Figure 3-9 Adjust the Endpoints setting so that they'll appear on the model.

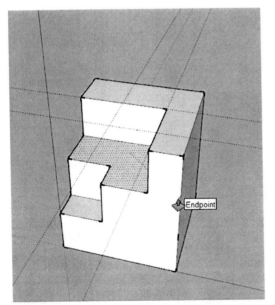

Figure 3-10 Push/pull the top planes down and inference match them to the division endpoints.

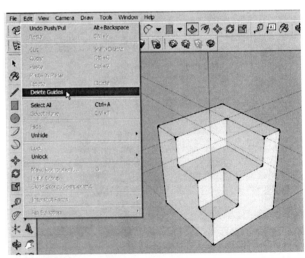

Figure 3-11 Delete all guidelines at once at Edit > Delete Guides.

Name Stand Project

Let's make a cool name stand to put on your desk (Figure 3-12). In the process, you'll see how to use the 3D Text, Move, and Follow Me tools and learn another trick with the Tape Measure tool.

Before starting, model a small block. This is useful on many projects to help orient some tools, as we'll see shortly. Place the block near the origin and above the red and green axes.

Figure 3-13　The 3D Text tool.

Figure 3-12　Name stand.

1. Model your name. Activate the 3D Text tool (Figure 3-13) located on the Large Tool Set. A dialog box appears: type your name, or my name if you want to follow this tutorial along exactly as is. If the text won't stand

straight up, run the Text tool over a vertical face on the cube (Figure 3-14). It will orient to that face. Then click to place. You can't edit this text once you click it, at least without an editing extension. So if you want to change it later, delete and redo it.

The text you just typed is a *component.* When you click on it, it will highlight like a group. And like a group, its geometry is inside an invisible shell, meaning that it won't stick to other geometry and must be double-clicked to edit. However, when you make copies of a component, editing one edits them all.

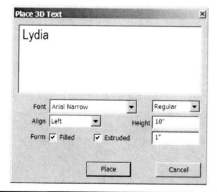

Figure 3-14　Type your name with the 3D Text tool. Run the tool over a vertical face of the cube to help orient the text vertically.

2. Make the stand (Figure 3-15). With the Rectangle tool, make a rectangle slightly bigger than the text. You can eyeball the size or type it, but in this case, type **2'11,1'** to make a rectangle 2'11" long and 1' wide. Remember, there are no spaces between any of the characters. The reason for this overly large size is that it's similar to the default size of the 3D text. We'll scale the model down to an appropriate size when we're finished. The space in between the rectangle's edges should darken, indicating that it's a face. If it doesn't darken or highlight when you try to select it, there's no

face. Push/pull the face up 1/2" (type **1/2** or **.5**) to give it some thickness.

3. Bevel the stand. Draw a line with the Pencil tool as shown in Figure 3-16. Then activate the Follow Me tool (Figure 3-17). Like Push/Pull, this tool extrudes a face into a form, but it can extrude around corners, not just up and down.

Click the Select tool on the bottom of the stand. This will make its perimeter the extrusion path. Then activate Follow Me. The bottom will deselect, but that's okay. Click on the triangular face shown in the

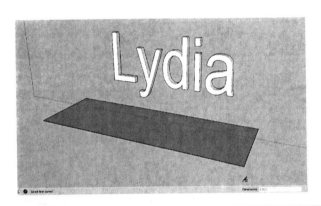
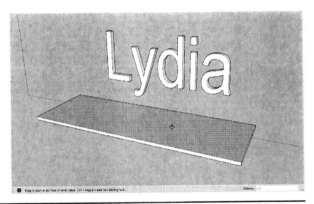

Figure 3-15   Use the Rectangle tool to make a face; then give it a thickness.

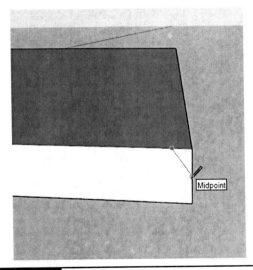

Figure 3-16   Draw a line on a corner.

Figure 3-17   The Follow Me tool.

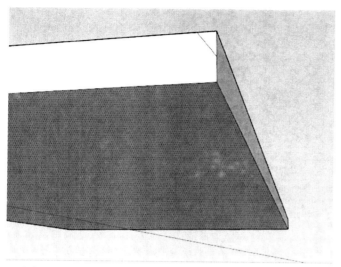

1. Select the bottom.

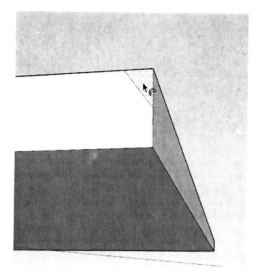

2. Click Follow Me on the profile.

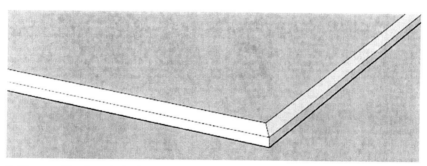

3. The resulting beveled edge.

Figure 3-18 Click the Select tool on the bottom of the stand. Then click Follow Me onto the profile to create a bevel.

second graphic in Figure 3-18. Follow Me will race along the path and make a bevel.

4. Move the text to the stand. First, make the stand a group so it's easier to work with. Triple-click the Select tool on the stand, right-click, and choose Make Group (Figure 3-19). Next, click the Select tool onto the text to highlight it.

Click on the Move tool (Figure 3-20). It relocates selected geometry. Move is self-selecting, meaning that it highlights geometry when it touches it. However, if the geometry doesn't select (this sometimes happens), just click the Select tool on it, and

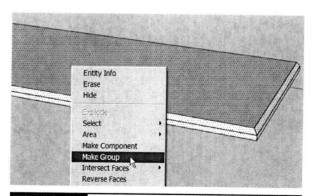

Figure 3-19 Make the stand a group.

Figure 3-20 The Move tool.

then click Move. Click Move onto a bottom edge of one of the letters. Watch for a pop-up inference saying that you've clicked an edge, and move it to the stand. Watch for another inference saying that the text is indeed on the stand's face (Figure 3-21). Note the four small crosses you see when moving the model, those are pivot (rotation) points. If you click on them, you can rotate the model around those points instead of moving it linearly.

Figure 3-22 Click the Editing box open and move the "y" up.

Figure 3-21 Move the text to the stand.

5. Position letters as needed (Figure 3-22). The tail of the "y" is below the stand and needs to be moved up. Double-click the Select tool on the text to open its Editing box. Note that everything outside the box turns gray, meaning that it's uneditable. Orbit under the stand, click Move on a bottom edge of the tail, and move the letter straight up—watch for a blue inference line indicating that you are indeed moving straight up—and click it onto the top face of the stand. Remember,

whenever you move geometry, move it along the axes unless you have a specific reason not to. Keeping geometry aligned with the axes makes it easier to work with.

6. Scale the model with the Tape Measure (Figure 3-23). At 2'11", this model is too big. Let's make it 12" long. With the Select tool, drag a window around the text and base to highlight them. Then click the Tape Measure on the base's left and right sides. With the second click, a pop-up will appear telling you the distance between those clicks. Immediately type **12**. Remember that you don't need to type the inch sign because inches are the default. A dialog box appears asking if you want to resize the model. Click "Yes," and the model will scale down to 12" long.

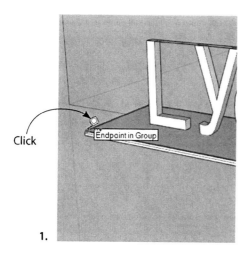

Click → Endpoint in Group

1.

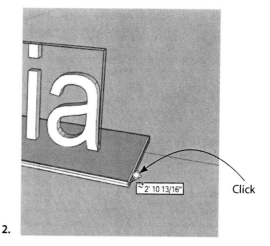

~2' 10 13/16"   ← Click

2.

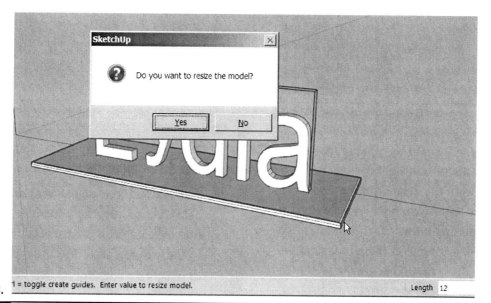

SketchUp

Do you want to resize the model?

[Yes]   [No]

1 = toggle create guides. Enter value to resize model.   Length 12

3.

Figure 3-23 Scale the model with the Tape Measure.

You have to type the new measurement immediately after the second click. If you do anything in between, it won't resize. Everything in the workspace will get scaled. To limit scaling to just one item, it must be a group or component. Click open its editing box and scale it inside that box. Know that this will affect all copies made of it. If you don't want a component to be affected, right-click it and select Make Unique. Scaling a component outside its editing box will only affect that component.

Tip: SketchUp becomes glitchy with very small (~1" long or less), or very big (~1,000' long or more) models and with dimension precisions set to less than 1/16". Common glitches include *clipping*, which is part of the model disappearing when you orbit, and axes inferences not appearing. Deal with this by enlarging or reducing the model's size while it's under construction. Thus you might multiply the dimensions of a tiny model by 1,000 or divide a giant one by that amount to keep the math easy and then scale it to the proper size when finished.

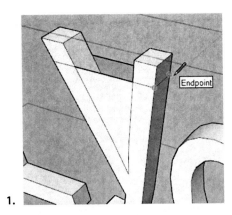
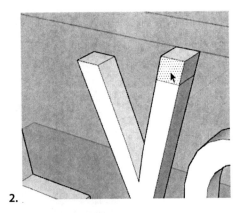

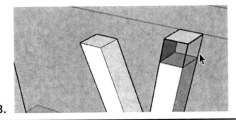

Figure 3-24 Shorten the top of the "y" by drawing lines on it and deleting everything above the lines.

7. Shorten the top of the "y" (Figure 3-24). It looks a little tall, so we'll delete a portion at the top. With the Pencil, trace a line around the whole letter. When you draw a line on a face, you break the face. It becomes two faces, and you can edit them independently. Click on the faces and edges above the line you just drew, and hit the DELETE key. Remember that if you hold the SHIFT key down while selecting them, you can select and delete all of them at once.

8. Add a stand to the dot over the "i" (Figure 3-25). Obviously, the dot won't suspend like that in a 3D print, so we need to attach it to the lower part. Orbit under the dot and draw a rectangle on it. Then Push/Pull the rectangle down until it touches the lower part (watch for an inference). Done!

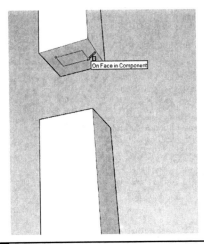
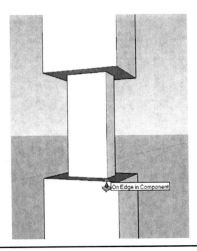

Figure 3-25 Add a stand to the dot over the "i."

Chain-Link Bracelet Project

Everyone likes a nice bracelet! We'll make the one in Figure 3-26 with the Circle, Scale, Rotate, and Follow Me tools. We'll also learn how to radial array, download a model from the 3D Warehouse, create a component, hide geometry, and use the Outliner.

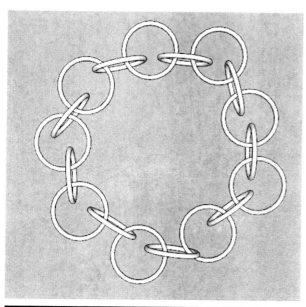

Figure 3-26 A chain-link bracelet.

1. Make a circle. Click on the Circle tool (Figure 3-27). Note that the number 24 appears in the Measurements box (Figure 3-28). That's because a circle is actually

Figure 3-27 The Circle tool.

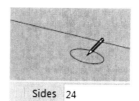

Sides 24

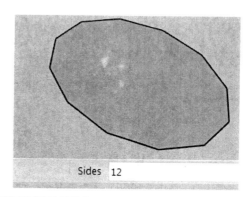

Sides 12

Figure 3-28 A circle is actually a polygon with 24 sides by default. You can adjust that number up or down.

You can make a cylinder (a circle that is pushed/pulled up) smoother by grouping it, right-clicking, and choosing Smooth/Soften. A slider will appear to adjust the amount of smoothness (Figure 3-29).

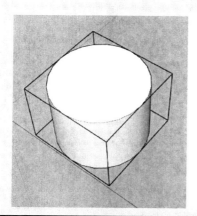

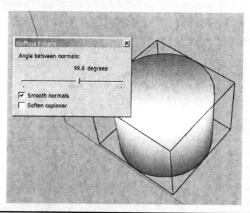

Figure 3-29 Soften a cylinder by right-clicking on it and choosing Smooth/Soften.

a collection of straight lines, 24 of them by default. A circular item will look like the polygon it is when 3D-printed. Type a smaller number and hit ENTER to make the polygon shape more obvious. Type a larger number, and hit ENTER to make both the shape and resulting print smoother. Whatever number you type becomes the new default. Note that the Circle tool has a drop-down arrow for accessing other tools.

Now click on the workspace to set the circle's center, type **1** to make a 1" radius, and press ENTER. The circle becomes a plane; that is, a face. You can change the

radius at this point by typing a new number before doing anything else. Once you perform another action, the size of the circle must be changed with other methods.

2. Model a torus as the first chain link. We'll make it by extruding a second, smaller circle around the larger circle's perimeter. Make a little cube and click the Circle tool onto a vertical face to orient it vertically. Then move it to the first circle, holding the SHIFT key down to maintain the vertical orientation (Figure 3-30). Click it onto the first circle, type **1/8**, and press ENTER.

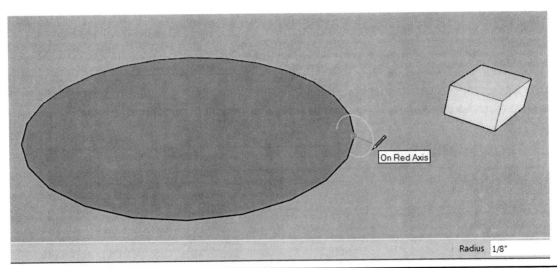

Figure 3-30   Click a second, smaller circle onto the first circle. Hold the SHIFT key down to maintain its vertical orientation.

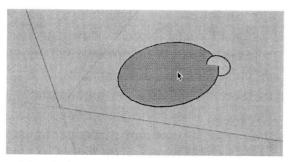

1. Select the large circle.

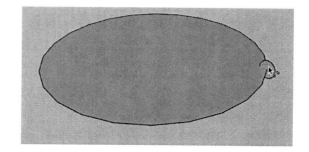

2. Click Follow Me on the small circle.

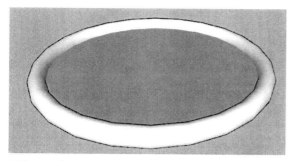

3. The small circle extrudes around the large circle.

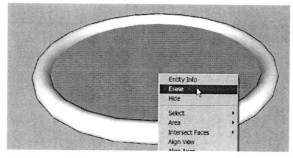

4. Select and delete the large circle's face.

Figure 3-31 Extrude the small circle around the large circle, and then delete the large one.

Click on the first circle to select it. Click on Follow Me, and then click on the small circle to extrude it around the large one (Figure 3-31). Then highlight the large circle, right-click, and choose Erase (or press the DELETE key).

3. Group and copy the chain link (Figure 3-32). Double-click on the link to select it, and click on Edit > Make Group. Then copy it by activating the Move tool and pressing the CTRL key. Drag the copy along the red axis and click to place it.

4. Rotate one link. Click on the Rotate tool (Figure 3-33). Like the Circle tool, you can run it over a cube's vertical face to orient it. The cursor color coordinates with the axis around which you're rotating the link. A black rotator means that the link isn't

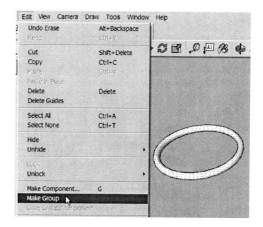

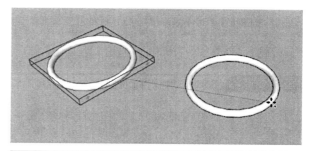

Figure 3-32 Group the link and copy it by activating Move and pressing the CTRL key.

Figure 3-33 The Rotate tool.

aligned with any axis. The Rotate tool also will cycle through the orientations if you hold the left mouse button down and drag the cursor around. All orientations are most likely to appear if you model near the origin. Hold the SHIFT key down to lock the orientation, and move the cursor to the link. Click on one end of the link to set the pivot, and then click on the other end. Rotate the link until it snaps to the blue axis (Figure 3-34).

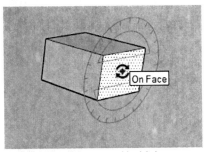

1. Orient the Rotator and hold the SHIFT key down to lock the orientation in place.

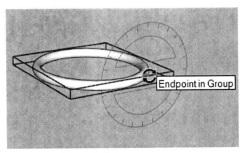

2. Click on the end of the group.

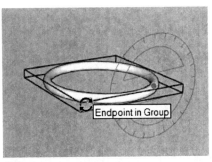

3. Click the other end of the group.

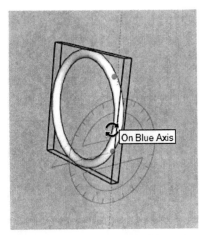

4. Snap the group along the blue axis.

Figure 3-34 Rotate one link parallel to the blue axis.

5. Move the two links together (Figure 3-35). Keep plenty of space between them; if they touch, they'll fuse together when 3D-printed. As an aside, when you move or rotate the links, you may notice that the original circle from which they were formed gets left behind. That's okay; just delete it.

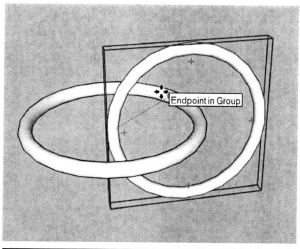

Figure 3-35 Use the Move tool to push the links together.

6. Make a component out of the two links (Figure 3-36). Select both by dragging a selection window around them. Then right-click on one and choose Make Component. A dialog box will appear with options. The only one relevant to this project is Replace Selection with Component. So check that box, name the component *Link*, and click Create.

7. Copy the component links around a circle via a radial array. *Array* means to copy in a specific pattern. Linear array copies items in a line; radial array copies items in a circle.

   a. First, make a 2½" radius (5" diameter) circle. Position the component as shown in Figure 3-37. You can use the Rotate tool or the pivots that appear with the Move tool.

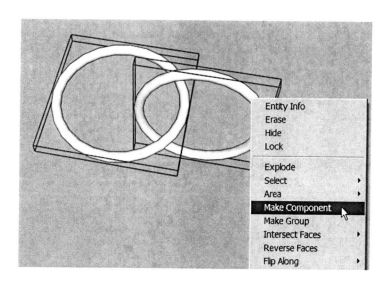

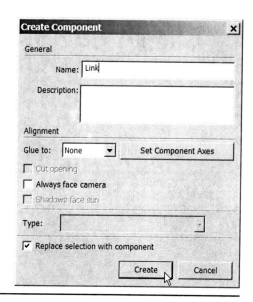

Figure 3-36 Turn the two links into one component.

Component options are:

- **Set the component axis.** Every component has its own axis. It sets the orientation upon insert and is the "handle" when moved. The default alignment is with the global axes, and the origin is in the corner of the bounding box closest to the global origin.

- **Set the alignment.** This applies to components that will be attached to faces, not stand-alone components. The alignment option determines which face the component "glues" or snaps to. If your component is a window, door, or piece of wall art, choose vertical to make it snap to walls. A vertical orientation also will make the component automatically rotate when moved to a perpendicular wall.

- **Cut opening.** This appears when any alignment setting besides None is chosen. It's relevant to door and window components to make them cut holes in the walls on which they're placed.

- **Set the component to always face the camera.** This is how the default scale figure operates. Have you noticed that no matter which way you orbit, it always faces you?

- **Replace selection with component.** Click this box. Otherwise, what you've selected will not be made into a component.

- **Always face camera box.** This is relevant when making 2D components such as the default scale figure. Check it or you'll see its unfinished backside when orbiting.

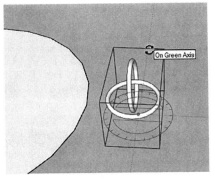

a. Make a 2½" radius circle and position the component links as shown.

Figure 3-37 Make a 2½" radius circle, and rotate the component as shown.

b. Select the component, activate the Rotate tool, and move the cursor to the center of the circle (Figure 3-38). It should snap to a center inference; when it does, click it into place. If it doesn't snap to a center inference, encourage the inference to appear by hovering the mouse over the circle's perimeter, and moving it slowly toward the center. Click a second time onto the component links. Then press and release the CTRL

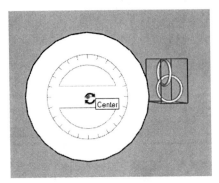

b. Select the component and click the Rotate tool onto the circle's center.

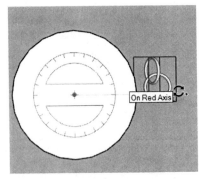

Press and release the CTRL key. Then click the Rotate tool onto the component.

Figure 3-38 Select the component, activate Rotate, click the cursor to the circle's center, click it on the component, and click the CTRL key.

key. A plus sign (+) appears, signifying that multiple copies will be made.

c. Move the cursor down; one copy will appear. Don't click it into place; instead, type **360**, and hit the ENTER key (Figure 3-39).

d. Type **/6**. Five copies will array around the circle; the sixth is the original component (Figure 3-40).

8. Move the components together to connect the links (Figure 3-41). You'll need to rotate them in the process.

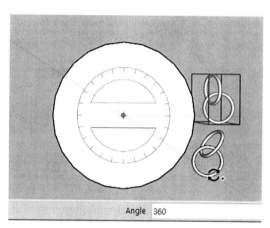

c. Move the cursor down and type 360.

Figure 3-39 Move the cursor down, and type **360**.

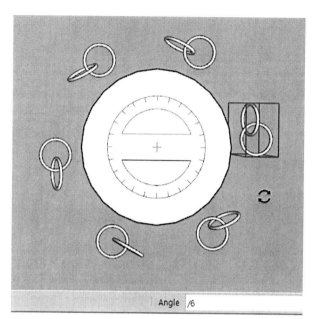

d. Type /6.

Figure 3-40 Type **/6** to array five copies around the circle.

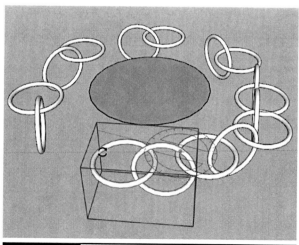

Figure 3-41 Move the components together to form a connected bracelet.

Here are some tips while moving the links:

- If the circle is in your way but you don't want to erase it, hide it by selecting it, right-clicking, and choosing Hide (Figure 3-42). Unhide it at Edit > Unhide.

- To edit one component but not its copies, select that component (don't open its editing box), right-click, and choose Make Unique (Figure 3-43).

- Click on Window > Outliner to see a hierarchical tree of all components and groups (Figure 3-44). Click on an entry

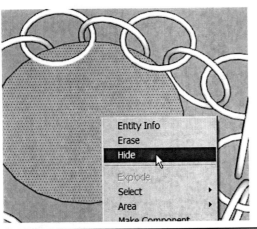

Figure 3-42   Hide geometry to get it out of the way.

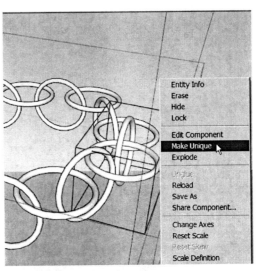

Figure 3-43   Make a component unique so that edits don't affect copies of it.

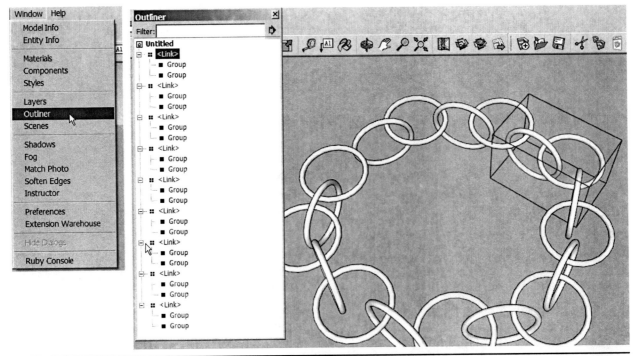

Figure 3-44   The Outliner shows a hierarchical tree of groups and components. This is a hierarchy outline of the bracelet.

in the outliner to select it in the model. This is useful when there are nested groups and components, as it removes the need to drill down through them all to reach a specific one. Figure 3-44 is the outline of the bracelet; you can see that each link is a separate group, and two links form one component.

9. Scale the bracelet if needed. A visual of it on a 3D scale figure would tell us if it needs to be scaled up or down. Click on the Warehouse icon (Figure 3-45). This takes us to a magical place called the *Trimble 3D Warehouse*, a repository of millions of models (Figure 3-46). You need Safari or Internet Explorer to access it this way, that is, from inside SketchUp. Any browser can access it directly on the Web at https://3dwarehouse.sketchup.com.

A search for "3D figure" returned the model shown in Figure 3-47. I downloaded it and moved the bracelet onto its wrist, rotating it straight with the pivots that appear with the Move tool. Looks big!

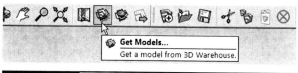

Figure 3-45 The Warehouse icon.

Figure 3-46 The Trimble 3D Warehouse has millions of models and a search box at the top.

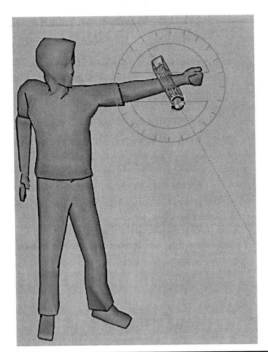

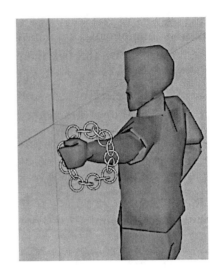

Figure 3-47   Move the bracelet onto a 3D figure to check its scale. Here it's being rotated with the Move tool's pivots.

Figure 3-48   The Scale tool.

Now we need the Scale tool (Figure 3-48).

10. Click it onto the bracelet. Green cubes called *grips* (Figure 3-49) appear. Hover the mouse over a corner grip. It will turn red to show that it's activated. Grab that red grip and press the CTRL key, which will force scaling around the center of the selection. Then drag the bracelet smaller until it's a more appropriate size.

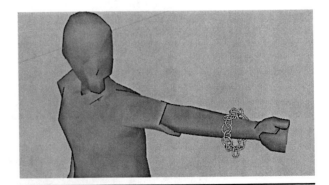

Figure 3-49   Scale the bracelet down.

The Scale tool resizes a whole model or part of it. You can "eyeball" sizes until they look right or type numbers for precision. For example, to adjust the bracelet to 5" long, triple-click on it with the selection tool to highlight it all. Then activate Scale. Grips will appear on all sides. Grab one grip, drag it a random length in any direction, and then immediately type **5,5,5**. The bracelet will adjust to that size. Typing **0.5** will scale it down to half its size. If you want to scale it in a different unit—such as 12 meters or 12 millimeters—type **12m** or **12mm**.

Different grips create different scaling effects. Corner grips scale proportionally (as does holding down the SHIFT key while holding any grip). Edge and side grips distort geometry. To change a rectangle's size, activate Scale, grab a grip, randomly move it, and type two numbers separated by a comma. The first number scales along the red axis, and the second number scales along the green axis.

Pet Collar Tag Project

Our favorite pet needs a shiny new tag. We're going to make the one in Figure 3-50 by importing and tracing a picture. In the process we'll use the Eraser, Two Point Arc, and Freehand tools; mirror geometry; change line color and thickness; flip inverted faces; and repair holes.

1. Import the picture (Figure 3-51). Click on File > Import to bring up a navigation

Figure 3-50 A pet collar tag.

 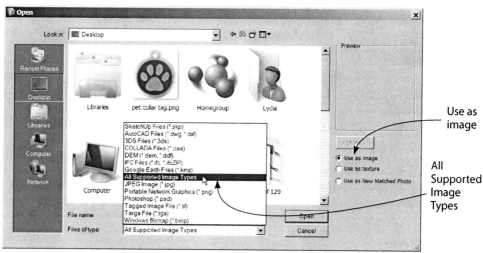

Use as image

All Supported Image Types

Figure 3-51 Import the picture.

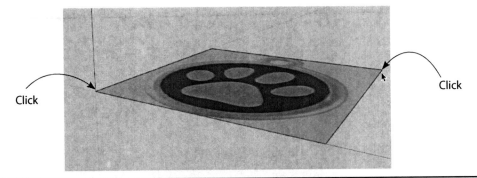

Figure 3-52 Click two corners to size and place the picture.

window. Click on the Use As Image button, and set the Files of Type field at the bottom to All Supported Image Types. Locate and click on the picture and then click Open. Click to place the first corner, drag the image larger, and click to place the second corner (Figure 3-52). If you accidentally let go of the cursor and the image becomes too small, just undo and reimport it. We'll scale it later.

SketchUp Make/PC imports .jpg, .png, .tif, and .bmp raster files. SketchUp Make/Mac imports the same plus .pdf files. .Gif files are not importable. SketchUp Pro imports the files shown in Figure 3-51. SketchUp resamples (downsizes) files larger than 1,024 × 1,024 pixels. However, that size is much larger than what's usually needed. Because large files slow down the software, crop and resample them yourself before importing.

2. Trace the tag's body. First, click the Plan icon on the Views toolbar because it will make tracing easier (Figure 3-53). Then click the Circle tool in the middle of the tag and make two circles as shown in

Figure 3-54. Be aware that it's unlikely that you'll be able to trace any image exactly because there's nothing to precisely click on; everything is eyeballed. Plus, SketchUp's tools may not be precise enough to trace intricate pictures. However, make sure that you see the On Face in Image inference each time you click on the picture because that will keep your points coplanar (all in the same plane). If the endpoints are not coplanar, faces will not form.

Figure 3-53 View the picture in Plan.

1.

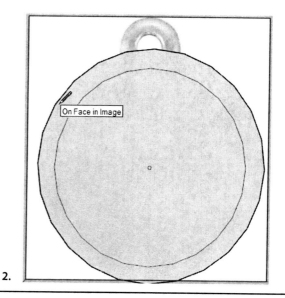

2.

Figure 3-54 Trace the tag's two large circles.

Tip: Modeling parallel to the axes and diligently watching for inferences are the two biggest things you can do to make a good model. A model with many noncoplanar faces is difficult or impossible to fix; scrapping it and making a new model usually takes less time.

3. Make the faces transparent (Figure 3-55). The circles' faces obscure the picture, but we can fix that. Go to View > Face Style, and click X-ray. Now we can see the picture again.

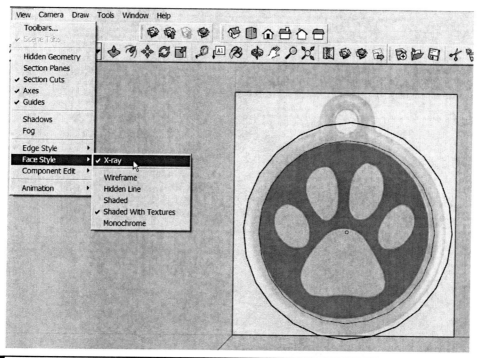

Figure 3-55 Make the faces transparent to see the picture.

4. Trace the tag's hole (Figure 3-56). Make another two circles. Again, watch for the On Face inference. Then use the Eraser tool (Figure 3-57) to make the hole look like Figure 3-58. The Eraser only erases edges, not faces. Click it onto an edge to remove it. You also can drag the Eraser by holding the mouse key down to continuously erase multiple edges.

5. Trace the paw. We won't be able to trace it exactly, but we'll do what we can.

   a. As per Figure 3-59, click the Circle tool onto a pad and copy it once (remember, you can do this with MOVE + CTRL or CTRL + C and CTRL + V). Warp the circles by clicking the Scale tool onto them and moving the grips. Then click the Rotate tool onto them to adjust their positions.

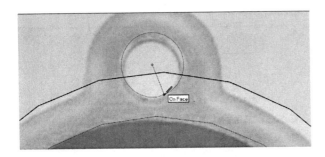

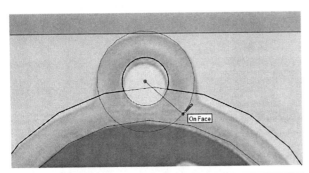

Figure 3-56 Trace the hole's two circles.

Figure 3-57 The Eraser tool.

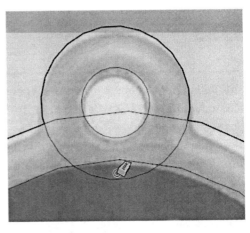

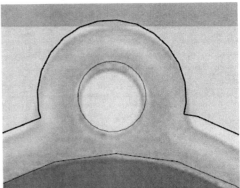

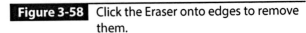

Figure 3-58 Click the Eraser onto edges to remove them.

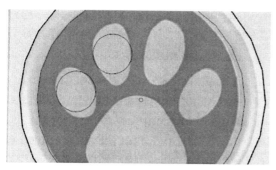

1. Draw two circles.

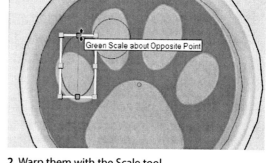

2. Warp them with the Scale tool.

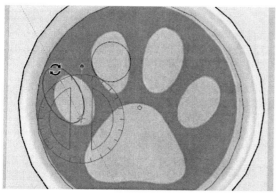

3. Rotate them.

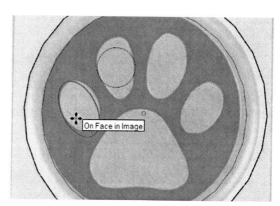

4. Copy and miror them.

Figure 3-59   Make two pads with the Circle, Scale, and Rotate tools.

b. Mirror the pads (Figure 3-60). First, copy them. To do this, select them (remember, you can do this by dragging a selection window around them, or you can double-click one, hold the SHIFT key, and double-click the second one), activate Move, and press CTRL. Make sure that you move the copies along the red axis—watch for its inference.

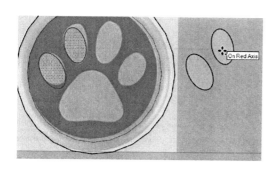

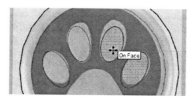

Figure 3-60   Mirror the pads by copying and flipping them.

c. With the pads highlighted, right-click and choose Flip Along > Red Direction. The pads will flip along the red axis. Then move them along the red axis into place.

d. Trace the big pad's round corners (Figure 3-61). Activate the Two Point Arc. Click the two endpoints (watch for the On Face inference), then drag the arc along the green axis (watch for the green axis color inference and the On Face inference). Click to set.

e. Trace the big pad's sides. Activate the Freehand tool (Figure 3-62). Click it on an arc endpoint (watch for the inference), and hold the cursor down to drag it to another endpoint (Figure 3-63).

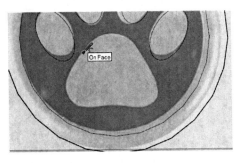
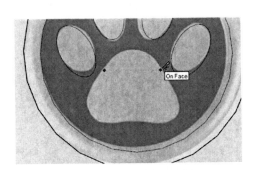
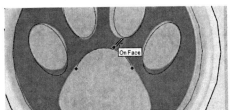
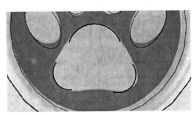

Figure 3-61 Trace the big pad's round corners with the Two Point Arc.

Figure 3-62 The Freehand tool.

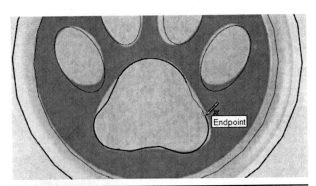

Figure 3-63 Drag the Freehand tool to make freehand lines.

6. Erase the imported picture by clicking the Eraser on it (Figure 3-64). If you think you'll need it later, hide it instead by selecting it, right-clicking on its perimeter, and choosing Hide. Group the tag by dragging a window around it to select it, right-clicking, and choosing Group.

7. Rotate the grouped tag upright (Figure 3-65). Model a cube, and run the Rotate tool over it to make the red cursor appear. Hold the SHIFT key down to lock that orientation. Then move it to the group, click at two points, and position it parallel to the blue axis. Watch for the On Blue Axis inference while rotating it upright.

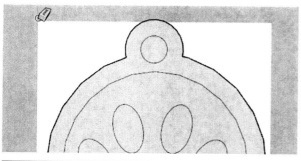

Figure 3-64 Erase the imported picture and group the tag.

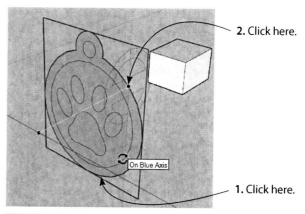

Figure 3-65 Rotate the grouped tag upright.

Tip: If the traced lines are hard to see, change how they're displayed. Go to Window > Styles, and click on the Edit tab. Check the Profiles box, and overtype its default number with a larger one. This will make the line display thicker. However, know that even with the larger profile number, some lines will still display thin. This is because SketchUp displays perimeter lines thicker than it displays lines embedded in a plane. To change the color, click on the black square in Edit box's lower-right corner. A dialog box will appear. Click on the color picker, move the slider up, and click OK to set the new color. To restore the line to black, move the slider all the way down (Figure 3-66).

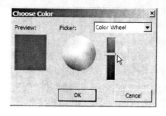

Figure 3-66 Change the color and thickness of the line display at Window > Styles.

8. Scale the tag to an appropriate size (Figure 3-67). Click the Tape Measure on the top and bottom of the tag. You need to click points on the tag; you can't click on empty space. Type **1 1/4** (put a space between the two 1s) to make the tag 1 1/4" tall. Click Yes when prompted, and the tag will rescale. It will become tiny, so click Zoom Extents to fill the screen with it.

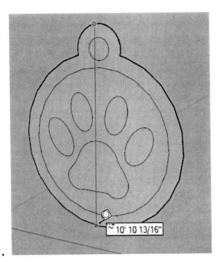

1.

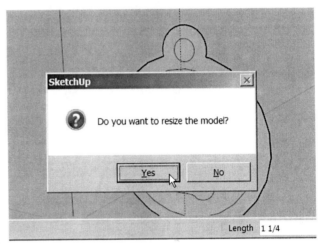

2.

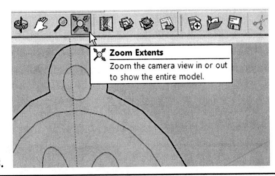

3.

Figure 3-67   Scale the tag down to make it 1¼" tall.

9. Add thickness (Figure 3-68). Double-click on the tag to open its group editing box. Then use Push/Pull to make the perimeter 1/8" thick. Remember to do all edits inside this box. If you edit outside it, you're really just drawing over it. Once you move the tag the edits will get left behind. Also, some tools won't work outside an editing box; Push/Pull will show a circle/slash symbol if you try to apply it outside a group. Next, push/pull the interior 1/16" thick. Finally, right-click on the face over the tag's hole, and delete it (Figure 3-69).

10. Reverse the faces, if needed (Figure 3-70). Push/pulling often causes faces to reverse. The front (white) side needs to face out. Select by holding the SHIFT key and clicking on geometry to add or delete from the selection. Holding the CTRL key down adds to the selection. Orbit around the tag to make sure that all faces and edges are selected. Then right-click and choose Reverse Faces.

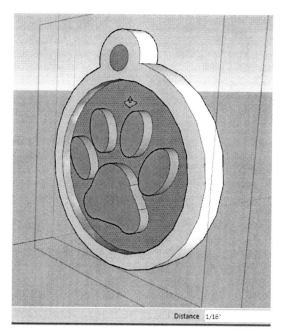

Figure 3-68 Push/pull the perimeter 1/8" thick and the face 1/16" thick.

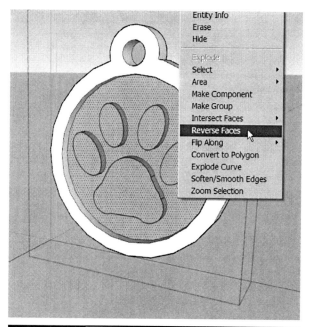

Figure 3-70 Select the face, right-click and choose Reverse Faces.

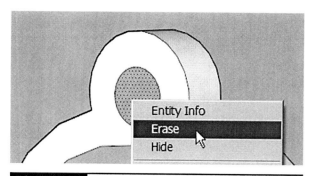

Figure 3-69 Delete the face over the hole.

11. Close the backside. Orbiting behind the tag, you can see the pad holes (Figure 3-71). Fill them in by tracing over one line on each pad with the Pencil (Figure 3-72). Don't erase the oval outlines that are left because that will mess up the front. Then trace over one line on the tag's body to close it (Figure 3-73). The resulting circle can be erased because it won't affect anything. Always erase unneeded lines to keep the model clean. This helps make a file 3D-printable.

Figure 3-71 The back of the tag has holes left from the push/pull operation.

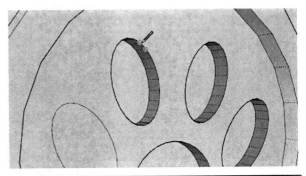

Figure 3-72 Close the holes by tracing one line on each.

1. Trace one line.

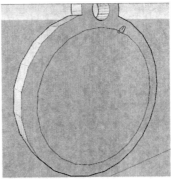

2. Erase the circle.

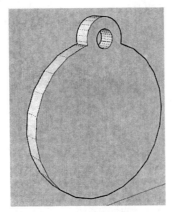

3. Final smooth back.

Figure 3-73 Close the back by tracing one line on the hole and erasing the resulting circle.

Coffee Mug Project

Let's make the coffee mug in Figure 3-74. In the process we'll use the Offset tool, Intersect with Model function, learn how to taper, access the Entity Info box, change a circle's radius, smooth edges, and set a parallel projection view.

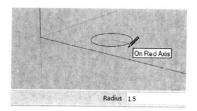

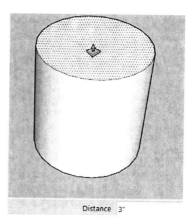

Figure 3-75 Make a cylinder.

Figure 3-74 Coffee mug.

1. Make a cylinder (Figure 3-75). Draw a 1 1/2" radius circle, and push/pull it up to 3".

2. Hollow the cylinder. First, activate the Offset tool (Figure 3-76). Then click it onto the perimeter edge, drag it inward, then type **1/4**, and hit ENTER. The cylinder will snap to that distance. Then push/pull the center down a bit and type **2 1/4** (Figure 3-77).

Figure 3-76 The Offset tool.

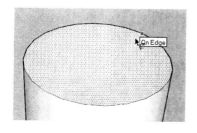

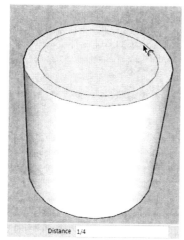

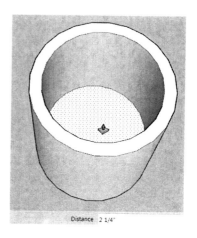

Figure 3-77 Hollow the cylinder with Offset and push/pull.

Tip: Use the Offset tool to make any model hollow. For instance, if you've modeled a house and only want to show the exterior, offset the walls to make the model hollow inside. This makes it cheaper and faster to print because it requires less material.

Taper the cylinder. Do this by double-clicking on the top to select its face and edges. Click the Scale tool and drag the corner grips while holding the CTRL key to scale about the center. Watch the Measurements box and click ENTER when you see 1.50 (Figure 3-78).

3. Adjust the thickness (Figure 3-79). The mug looks a bit too thick, doesn't it? Highlight the inner circle, right-click, and choose Entity Info. This brings up a box with information about the selected geometry.

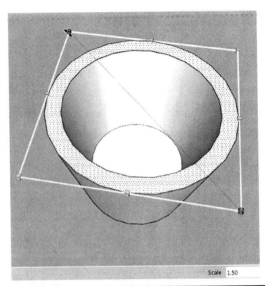

Figure 3-78 Taper the cylinder with the Scale tool.

Different options appear based on what's selected. Here we can overtype the existing radius with a new one. Type **2 1/8** and the mug's interior will adjust.

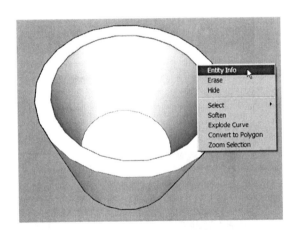

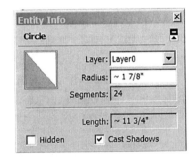

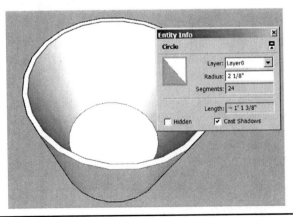

Figure 3-79 Change the inner circle's radius via its Entity Info box.

4. Smooth the mug's top edge (Figure 3-80). Select both circles (hold the SHIFT key down to do this), right-click on one, and choose Soften/Smooth Edges. Then move the slider until the edges look how you want them to. We'll discuss smoothing curves more later in this chapter.

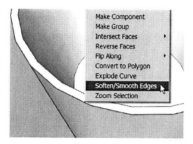

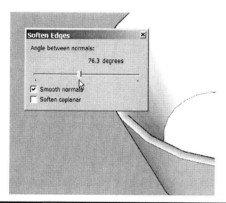

Figure 3-80 Smooth the mug's top.

5. Make a handle. First, group the cup so that nothing sticks to it (drag a selection window around it, right-click, and choose Make Group). Then follow these steps:

   a. Turn on Parallel Projection mode (Figure 3-81). When you click on the Views toolbar's Top icon, you get an aerial view, meaning that it's still 3D. To make it display as a true plan view (2D), click on Camera > Parallel Projection. This mode makes aligning objects easier.

   b. Model a cube (Figure 3-82). Put it next to the mug, and inference match its height to the mug. Do this by pushing/pulling a rectangle up, and before letting the cursor go, drag it to the mug. When

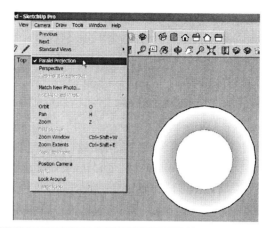

Figure 3-81 Turn on Parallel Projection mode to display the model in 2D.

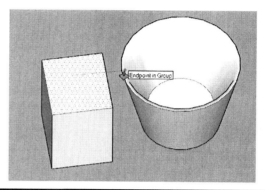

Figure 3-82 Model a cube the same height as the mug.

an inference symbol appears, let go. The cube's height will match the mug's.

   c. Click the Views toolbar's Front icon, and then group the cube so that nothing sticks to it (Figure 3-83).

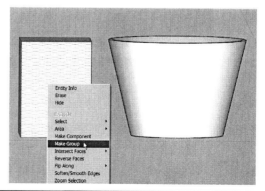

Figure 3-83 Display the model as a front view, and group the cube.

d. Use the Two Point Arc tool to draw an arc. Then warp it with the Scale tool (Figure 3-84).

e. Draw a circle perpendicular to the arc, snapping the center to the end of the arc (Figure 3-85).

f. Delete the cube. Then select the arc, activate Follow Me, and click on the circle to extrude it along the arc (Figure 3-86).

6. Position the handle on the mug (Figure 3-87). You'll probably need to switch between the plan and 3D views several times while doing this. Push the handle into the mug.

7. Delete the handle's protruding parts. Look closely at the intersection of the mug and handle, and you'll see that there is no edge between them (Figure 3-88). Double-click the handle to select it, then right-click it,

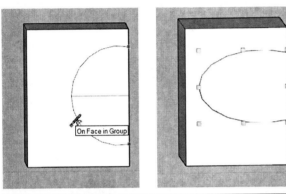

Figure 3-84 Draw and warp an arc on the grouped cube.

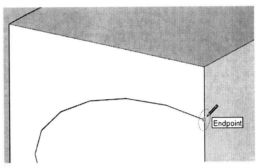

Figure 3-85 Draw a circle perpendicular to the arc.

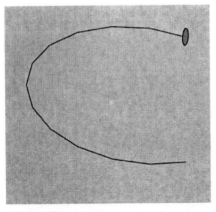

1. Select the arc.

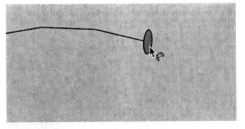

2. Click Follow Me on the circle.

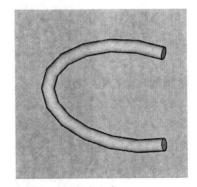

3. The extruded circle.

Figure 3-86 Extrude the circle along the arc.

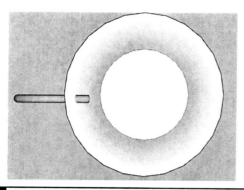

Figure 3-87 Position the handle on the mug.

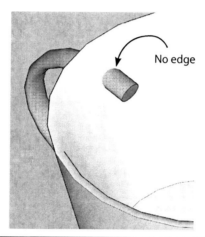

No edge

Figure 3-88 There are no edges between the intersection of the mug and handle.

Tip: Moving two pieces of geometry together is difficult when the pieces are off-axis and far from each other. Here's how to keep all pieces on the same axes. Draw guidelines by clicking the Tape Measure onto the axes and dragging them along the workspace (Figure 3-89). Then model all geometry along those guidelines. Remove guidelines individually with the Eraser or all at once at Edit > Delete Guides. Guidelines created inside a group or component must be erased from within the editing box.

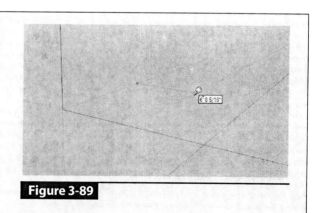

Figure 3-89

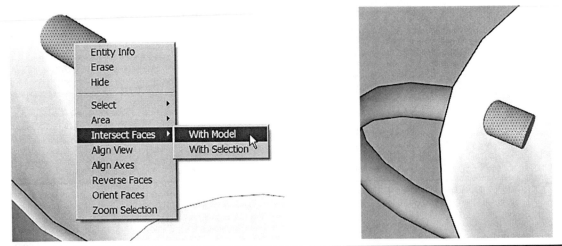

Figure 3-90 Create edges with the Intersect Faces function. Here, the protruding part can then be deleted.

and choose Intersect Faces > With Model. Edges will appear (Figure 3-90). Select and delete the protruding parts.

Pencil Holder Project

Here's something for Mom to treasure. We'll learn a technique called *autofold* to make the beveled letters in Figure 3-91.

1. Type **Mom**. Click on 3D Text, and choose a font (Figure 3-92). Simple, blocky ones are best; fonts with serifs and other flourishes do not 3D print well. Orient the text vertically with the help of a cube. Then explode it to make it easier to work with.

Figure 3-91 The beveled letters in this pencil holder were made by autofolding them.

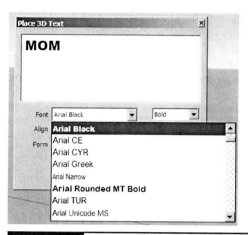

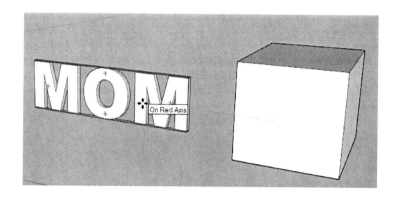

Figure 3-92 Choose a simple font, place the text, and explode it.

2. Make the letters 1" thick (Figure 3-93). Click the Tape Measure on an *M*'s back corner, move it forward, type **1**, and click. This makes a guide point 1" from the back corner that you can inference to. Push/pull the *M* to that guide point and then inference the other letters to it.

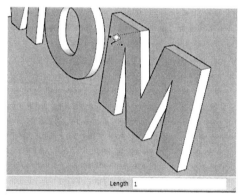

1. Make a guide point.

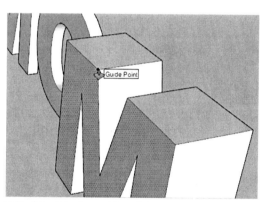

2. Push/pull to the guide point.

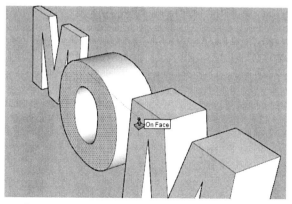

3. Inference match the letters' faces.

Figure 3-93 Make a guide point 1" from a back corner. Then push/pull and inference match the letters to it.

3. Offset the *M* (Figure 3-94). Click the Offset tool on the edge, move it in, and click to place. Different letters give different results; here some cleanup is needed.

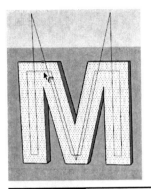

Figure 3-94   Offset the *M*.

4. Clean up the *M* (Figure 3-95). The offset *M* needs to be completely inside the larger *M* with no overlap. So you'll need to erase some lines and draw some lines, which can require some experimentation. For instance, erasing a line may cause faces to disappear. This can often be fixed by tracing one line with the Pencil. However, if tracing doesn't restore the faces, redraw the erased line, and try erasing a different line. Workflow order often makes all the difference in a good result. In this case, erasing the left line removed too many faces to fix. But erasing the right line removed just a few faces, which were easily fixed. Then I was able to erase the left line, and it didn't take any faces with it.

Drawing new lines may cause faces to appear that you didn't want. Just erase them (Figure 3-96).

Left Line   Right Line   Traced Line

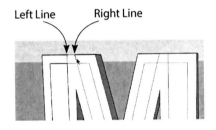

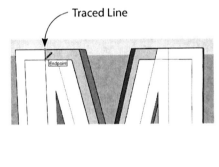

Figure 3-95   Erasing the left line removed too many faces, but erasing the right line only removed a few. Faces can often be repaired by tracing one line.

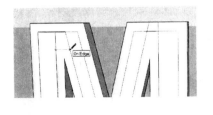

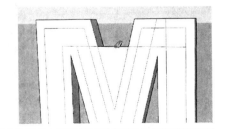

Figure 3-96   Unwanted faces may appear when drawing new lines. They can be erased.

1. Hightlight the offset face.

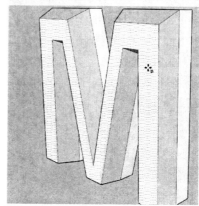

2. Autofold the face.

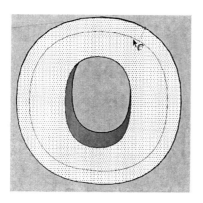

1. Offset the O.

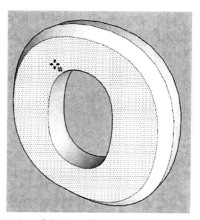

2. Autofold the offset face.

Figure 3-97   Autofold the *M* and *O* with MOVE + ALT (OPTION on a Mac).

5. Autofold the *M* (Figure 3-97). Highlight the offset. Then click Move onto the highlighted portion. Hold the ALT key down (Option on a Mac), and drag the offset forward. It will bevel. This can be tricky and does take some practice. Then offset and autofold the *O*.

6. Group and copy the *M*. Then delete the second *M*, and replace it with the first one. Group the *O* (Figure 3-98).

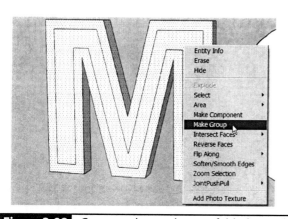

Figure 3-98   Group and copy the autofolded *M*. Delete the second *M*, and replace it with the copy.

7. Move all three letters so that they touch (Figure 3-99). Remember, they need to be grouped separately or they'll stick together.

8. Make the pencil holes (Figure 3-100). Draw a circle and push/pull it down. Select and copy it, and then move it to another location. If for some reason this doesn't work, you can also make one pencil hole, make the model transparent by clicking View > Face Style > X-ray, draw a second circle, and inference match the hole (Figure 3-101). Turn off X-ray by clicking it again.

Figure 3-99   Move the three letters together, and intersect their faces, if needed.

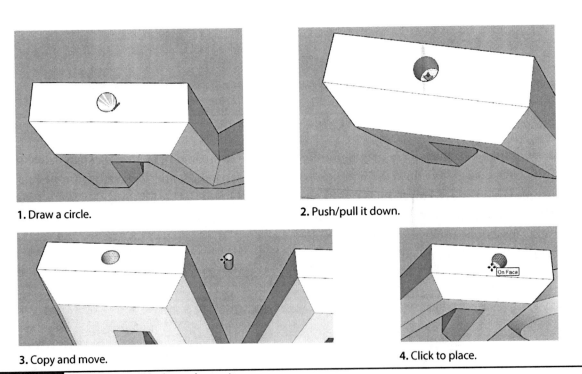

1. Draw a circle.

2. Push/pull it down.

3. Copy and move.

4. Click to place.

Figure 3-100   Make a pencil hole, and copy it.

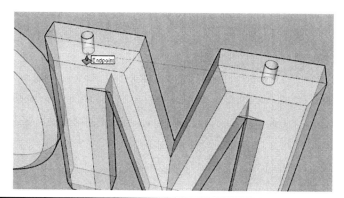

Figure 3-101 Make a pencil hole, draw a second circle, and with X-ray on, inference match the hole's depth.

Travel Mug

Let's draw the travel mug in Figure 3-102. We'll model the cup and lid separately using Follow Me, guide points, and the parallel inference.

1. Make a cube and group it (Figure 3-103). We'll use it as a surface on which to draw.

2. Draw the mug's 2D profile on the cube (Figure 3-104). Draw a 6" vertical line, a 2 1/2" horizontal line at the top, and a 1 1/2" horizontal line at the bottom. Connect the horizontal lines' endpoints, and then offset the profile by 1/8".

Figure 3-102 Travel mug.

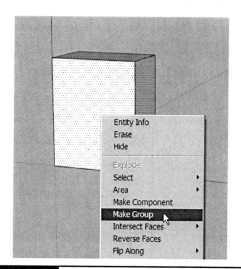

Figure 3-103 Make and group a cube.

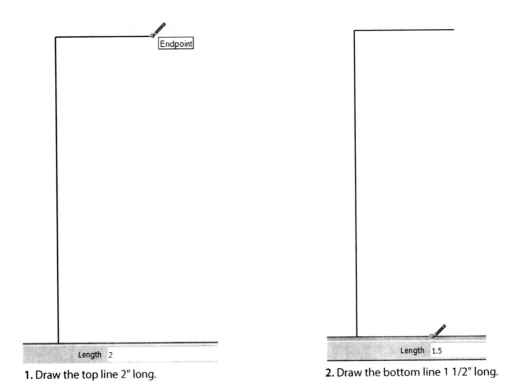

1. Draw the top line 2" long.

2. Draw the bottom line 1 1/2" long.

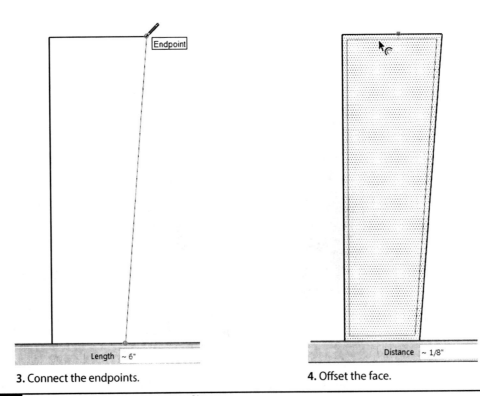

3. Connect the endpoints.

4. Offset the face.

Figure 3-104 Draw and offset the mug's profile.

3. Cut the profile in half (Figure 3-105). First, draw a line parallel to the profile's edge, as shown in Figure 3-105's first graphic. Encourage the Parallel to Edge inference to appear by touching the Pencil on the profile's edge and moving it slowly back to the left. When the Parallel to Edge inference appears, the line will be magenta. Then draw

a vertical line through the center of the whole profile. Delete everything to the left of that line.

4. Add the sleeve (Figure 3-106). Click the Tape Measure on the top of the profile, move it down, and type **1.5**. A guide point will appear at that location. Repeat, but type **2.5** instead. Then draw lines as shown

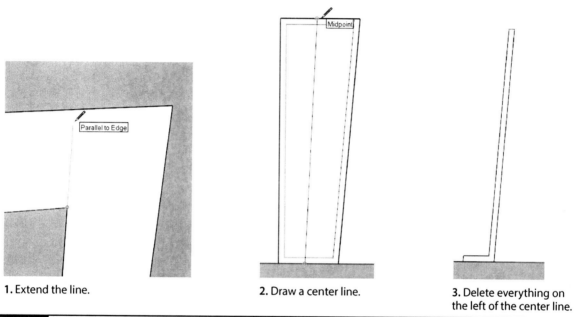

1. Extend the line.   2. Draw a center line.   3. Delete everything on the left of the center line.

Figure 3-105 Delete the left half of the profile.

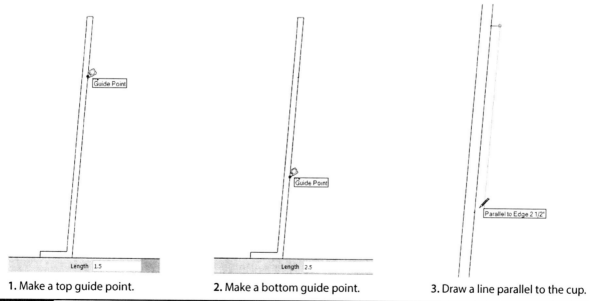

1. Make a top guide point.   2. Make a bottom guide point.   3. Draw a line parallel to the cup.

Figure 3-106 Make guide points for the sleeve's height, and then draw the sleeve using the Parallel to Edge inference.

in Figure 3-106's third graphic using the Parallel to Edge inference. When finished, the profile should have a face and no interior lines (Figure 3-107). You can verify whether a face exists by clicking on it to see if it highlights.

5. Turn the profile into a cup (Figure 3-108). Draw a circle at the bottom, locating its

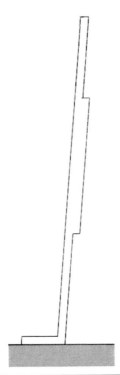

Figure 3-107 The finished profile should have a face and no interior lines.

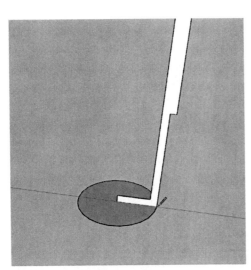

1. Rotate the profile around the circle.

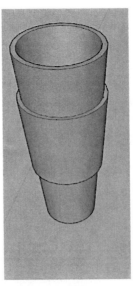

2. Rotated profile.

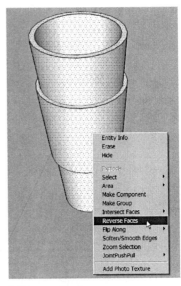

3. Reverse the faces.

Figure 3-108 Extrude the profile around a circle.

midpoint and radius as shown in Figure 3-108's first graphic. Select the circle, click on Follow Me, and then click on the profile. It will extrude into a cup. If the faces are reversed, select them, right-click, and choose Reverse Faces. And if the guide points inadvertently got extruded into dotted lines on the cup, click Edit > Delete Guides.

6. Fix the bottom (Figure 3-109). Orbit under the cup. The bottom probably will look like Figure 3-109's first graphic. Draw a line from one end of the circle to the other to form a face, and then erase the line. Orbit inside the cup, and push/pull the bottom up 1/8".

7. Draw a 4" circle for the lid (Figure 3-110). Then draw a larger circle slightly outside it and two circles inside it. Delete the original 4" circle when finished.

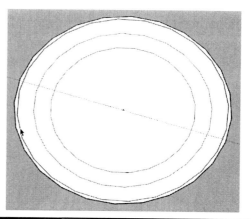

Figure 3-110 Draw the lid.

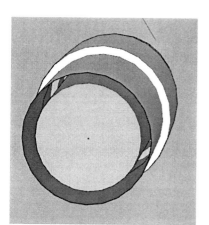

1. The broken bottom.

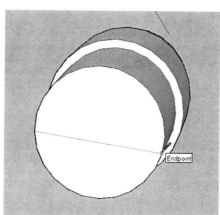

2. Draw a line to make a face, then erase the line.

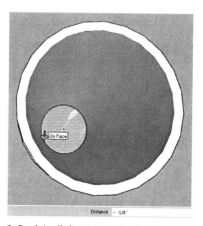

3. Push/pull the interior bottom up 1/8".

Figure 3-109 Fix the bottom and then push/pull it up.

8. Push/pull the rings up (Figure 3-111). Pull the first circle up 1/4" and the second circle up 1/2". Pull the third circle up so that it is just a bit below the second circle. Make sure that you pull them all up instead of pushing some down. This makes a difference when preparing it for 3D printing. In general, some workflows make a model *solid*, a necessity for 3D printing that we'll discuss in Chapter 5. Other workflows create models that aren't solid and difficult to make so, even if the appearance on the computer screen is the same.

9. Add the sipping hole (Figure 3-112). Draw a circle on the middle ring. Warp it with the Scale tool, and push/pull it down. Click the Push/Pull tool onto the middle ring's bottom edge to make a clean hole; otherwise, it will extrude past the bottom. If you have difficulty doing this and the hole keeps extruding past the bottom, just select the excess, right-click, choose Intersect

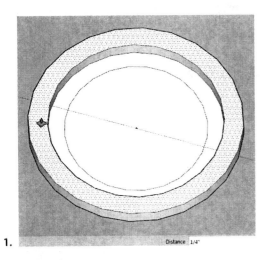

1.

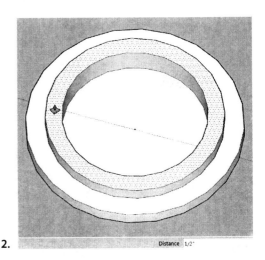

2.

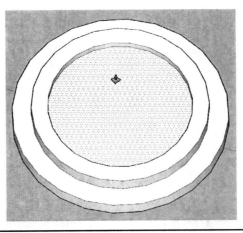

3.

Figure 3-111 Push/pull the rings up.

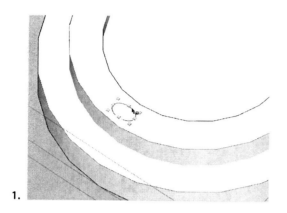

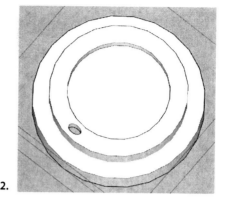

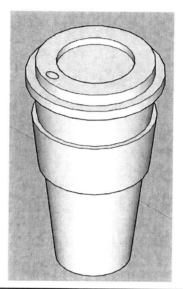

Figure 3-112 Add the sipping hole to finish the mug.

Faces with Model, and delete as needed. If a face remains, select and delete it, too.

Smoothing Out Curves

Let's look at an issue that curved features present. When a model has curved surfaces, the 3D print or CNC product will have lines that delineate the polygon edges, giving it a faceted appearance. You can't erase the lines or you'll remove the polygons themselves. But you can hide them. Select an edge, right-click on it and choose Hide. The edge will become invisible. You can bring it back at Edit > Unhide. Something else you can control for a smoother product are

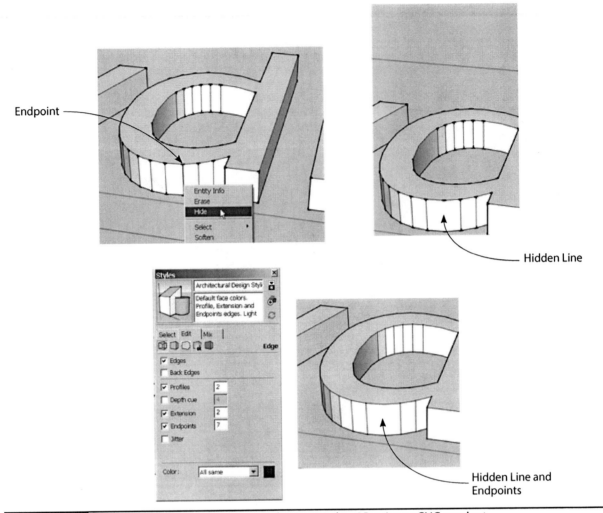

Figure 3-113   Hide lines and remove endpoints for a smoother 3D print or CNC product.

those "dots" all over the model (Figure 3-113). They're endpoints. Click on Window > Styles, click on the Edit tab, and uncheck Endpoints. They'll disappear, too.

tools. We scaled, intersected, offset, extruded, and mirrored them and turned them into recognizable objects. However, they are not yet 3D-printable; in Chapter 5 we'll learn how to make them so.

Summary

In this chapter we learned how to make multiple shapes and forms with SketchUp's native

CHAPTER 4

Projects Using SketchUp Pro and Extensions

In Chapter 3 we used SketchUp's native tools and made a component. In this chapter we'll download more tools called *extensions*. These enable you to perform operations the native ones can't. We'll also download and edit models from the Warehouse and another online site called Thingiverse and use SketchUp Pro's solid tools. This chapter's projects are a bench, a Halloween fingernail, a chocolate mold tray, an architectural terrain model, a Pac man ghost, and a floor plan.

Components

Access the Warehouse by clicking on Window > Components. A navigation browser will appear (Figure 4-1) showing two folders of component collections. The Sample collection is open by default. Click on the drop-down arrow to see more collections (Figure 4-2). You can drag and drop their components from their thumbnails into the workspace. Click on the House icon to see which components are currently in the model. Be aware that even after you delete a component from the model, it still remains in the file. Having many components contributes to a large file size, and that slows the model down. Get rid of unused components by clicking the House's fly-out arrow and then clicking Purge Unused (Figure 4-3).

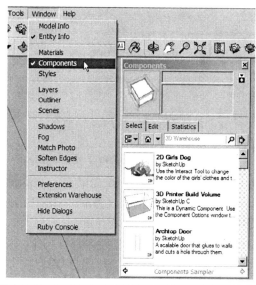

Figure 4-1   The Components browser with the Sample collection open.

Figure 4-2   Click the drop-down arrow to see more Component collections.

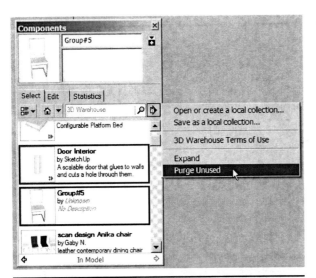

Figure 4-3 Purge unused components with the House icon's fly-out arrow.

Type what you're looking for in the Navigation browser's search field, and scroll through the thumbnail results. Click on one to download the component, and drag and drop it into the workspace. I downloaded the component shown in Figure 4-4.

This component consists of four chairs and a scale figure, each of which is a separate component. So we've got five components nested inside one large shell component.

I only want one chair, so double-click the editing box open and delete everything but one chair. Highlight that chair and explode it. This removes its outer shell, leaving just the component chair (Figures 4-5 and 4-6). Don't explode that.

This is an example of a simple edit. Warehouse models usually need some editing to make them suitable for your own purposes. Know that anyone can upload to the Warehouse, and there is no vetting. The SketchUp team uploads models, companies upload models of their products, and casual SketchUp users upload their creations. There are replicas of whole buildings and cities, and rooms filled with brand-name furniture, fixtures, and equipment. Excluding branding and logos, all content is free and can be used for both personal and commercial purposes. SketchUp also has a partnership with the online service bureau i.materialise.com; hence, you can access its large inventory of 3D-printable models through the Warehouse. Most user-uploaded Warehouse

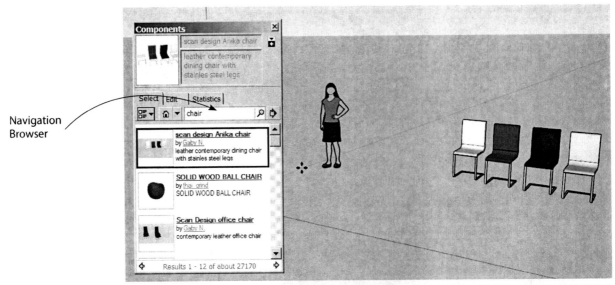

Navigation Browser

Figure 4-4 Download a Warehouse component via the Navigation browser's search field.

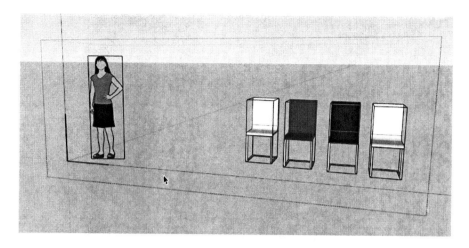

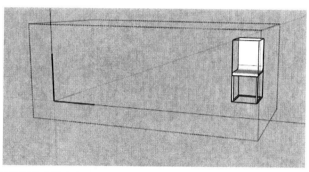

Figure 4-5   Open the component editing box, and delete unneeded items.

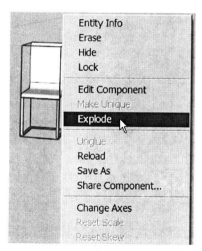

Figure 4-6   Select and explode the outer shell, leaving the remaining chair component intact.

components are not 3D-printable as-is, but many can be edited to be made so.

While many models are great, others are bloated (huge file size) or have other problems.

When you download a model, you download everything in it—imported images, layers, and possible corrupted elements. Before downloading a Warehouse model into a model that you've spent hours working on, download it into a new file first to check it out.

What's an Extension?

An extension, also called a *plug-in*, *script*, or *ruby* (named after its programming language), is a simple text file with an .rb or .rbz suffix at the end of the file name. It plugs into SketchUp to extend its native capabilities, much like a phone app. Extensions can bend 3D text, perform energy analyses, add light, and more. Most are written by third-party developers. Many are free, and others have a cost. They work with both Make and Pro.

The Extension Warehouse

The best source of extensions is the Extension Warehouse (EW). It contains both free and pay ones. Access it through its toolbar icon (top graphic, Figure 4-7).

At the top of the EW homepage is a search bar and login box. You can browse what's there without logging in, but you need to log in with a Google account to download anything. So log in, and then mouse over the login box to see its options. My Extensions accesses a list of every extension you've downloaded (Figure 4-8), which is useful when you want to return to an extension's page. The EW homepage displays the extensions that are most popular at the moment. You must have the IE or Safari browser installed

Figure 4-8 Once logged in, you can download extensions and access these options.

and set as default to access the EW from within SketchUp.

We're going to download seven extensions in this chapter: TT Lib^2, Bezier Curves Tool, CLF Shape Bender, Shell, SketchUp Stl, s4u Make Face, and Add Terrain Skirt. The first six are in the EW, and the last one is at sketchucation.com.

Figure 4-7 The Extension Warehouse icon and homepage.

Download the Extensions

Download TT Lib[2] first (Figure 4-9). Do a search for it, or if it's visible in the Popular Downloads box, click there to access its page. This extension is a collection of background utilities that some other extensions need to work.

When logged in, you'll see a red Install button on the extension page. Click it. You'll be asked for permission to install and notified when it's done (Figure 4-10). Because extensions are executable files, be careful about downloading them from other websites. Legit outside sources exist, but everything in the EW has been vetted by the SketchUp team. After you install TT Lib[2], the red button will say Uninstall.

Figure 4-9 The TT Lib[2] page.

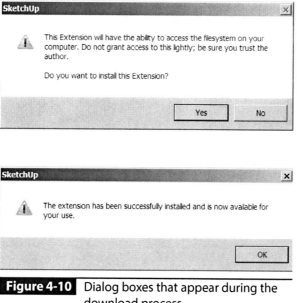

Figure 4-10 Dialog boxes that appear during the download process.

After downloading your first extension, an Extensions entry will appear in the menu at the top of the SketchUp screen. At Windows > Preferences > Extensions the downloaded extension will appear in a list (Figure 4-11). Note that there are some native extensions there, too. An extension's box must be checked for it to work.

Extensions that need TT Lib² to work will prompt you to download it if they can't find it on your computer. Some of these extensions don't work unless TT Lib² appears *above* them in the extensions list. So if you've already downloaded an extension that requires TT Lib², but you can't find that extension's tool—or it doesn't work—check its location in the list. If the extension is listed above TT Lib², uninstall it (but keep TT Lib²). Then reinstall it. It should appear listed below TT Lib² now, and the tool should work.

You can access the EW directly on the Web at https://extensions.sketchup.com, but downloaded extensions won't install automatically. In that case, and for extensions downloaded from other websites, click on the Install Extension button. If the Install Extension button doesn't work, drag the extension file or folder (don't remove a folder's contents) manually into the Plug-ins folder on your computer's program files. SketchUp 2015's path to this folder is Windows C:\Users\YOUR USERNAME\AppData\ Roaming\SketchUp\SketchUp 2015\SketchUp\ Plugins. Mac OS X users, go to User/Library/ Application Support/SketchUp 2015/SketchUp/ Plugins.

Search for and install Bezier Curves, CLF Shape Bender, Shell, SketchUp Stl, and s4u Make Face (Figure 4-12). The Bezier Curves extension draws French curves, which are noncircular arcs. The Shape Bender bends groups and components along a pre-drawn path. Shell applies thickness to a curved plane.

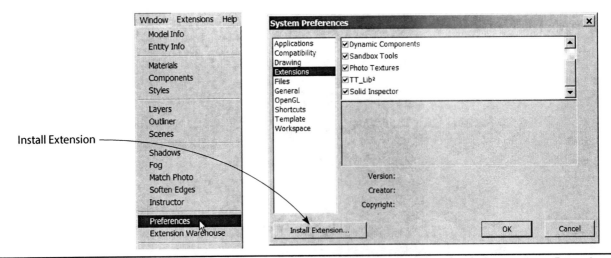

Figure 4-11 At Windows > Preferences > Extensions is a list of all native and downloaded extensions. Extensions that require TT Lib² to work need to appear listed below it.

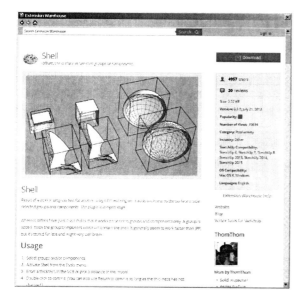

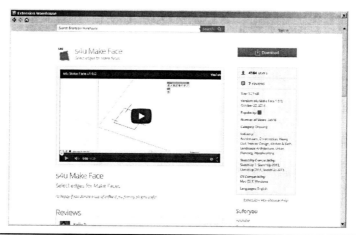

Figure 4-12   The Bezier Curves, CLF Shape Bender, Shell, SketchUp Stl, and s4u Make Face pages.

SketchUp Stl imports and exports a model as an .stl, the most popular file format for 3D printing. s4u Make Face creates a face inside a coplanar outline.

Each extension's page has a short video and instructions. They're written by the developers, and because some instructions are more informative than others, an extension may take some experimentation to figure out.

Where to Find the Extension Tools

Some extension tools are accessed through the Extensions menu at the top of the SketchUp workspace. Others appear as an entry in the Tools or Draw menu or as a submenu under one of them. Some appear in a context menu when the appropriate geometry is right-clicked. If the extension has a toolbar, it might appear in the workspace immediately on installation, You might have to check its box at View > Toolbars

first, or check its box in the Preferences > Extensions list. Or you might have to close and reopen the SketchUp file for it to appear. Let's look at the Bezier Curves extension now.

How to Use the Bezier Curves Tool

Draw and group a rectangle to serve as a drawing plane. Then click on the Draw menu to activate the Bezier Curves tool (Figure 4-13).

Drawing a Bezier arc takes four clicks (Figure 4-14). Click the two endpoints. Adjust the arc and click again. Click a fourth time to adjust the arc further and set it, or hold its position and click to set it. Watch for the On Face inference each time you click. To cancel the operation while drawing the arc, click Escape.

After the arc is set, you'll still be in the Bezier Curves tool. Click the Select tool to get out of it. To edit the curve, right-click on it and choose Edit Bezier Curve (Figure 4-15).

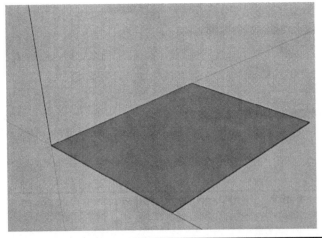
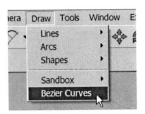

Figure 4-13   Draw and group a rectangle; then activate the Bezier Curves tool.

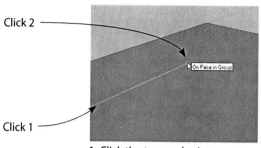

Click 2

Click 1

1. Click the two endpoints.

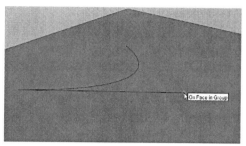

2. Adjust the arc and click.

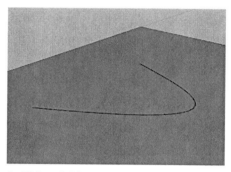

3. Click to finish.

Figure 4-14 Drawing a Bezier curve.

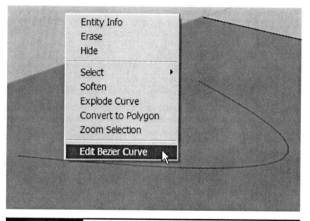

Figure 4-15 Right-click the Bezier curve to edit it.

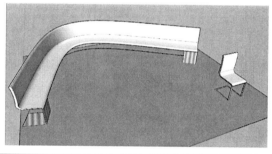

Figure 4-16 A curved bench made with a chair and two extensions.

Bench Project

We're going to use the Bezier Curves and CLF Shape Bender tools to turn the chair component we downloaded earlier into a long, curved bench that looks like Figure 4-16. The CLF Shape Bender tool only works on groups and components, not loose geometry.

1. Make a rectangle and group it. Then draw a Bezier curve on top of it (Figure 4-17).

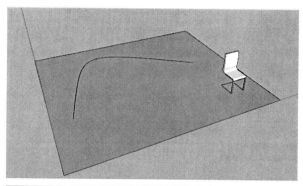

Figure 4-17 Make a rectangle, group it, and draw a Bezier curve.

2. Draw a line that's the width of the chair and runs along the red axis (Figure 4-18). Running along the red axis is critical, so watch for the red inference line.

3. At View > Toolbars, turn on the CTL Shape Bender toolbar (Figure 4-19).

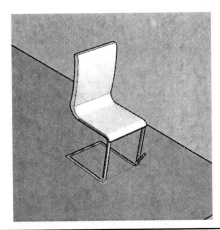

Figure 4-18 Draw a line that's the width of the chair and runs along the red axis.

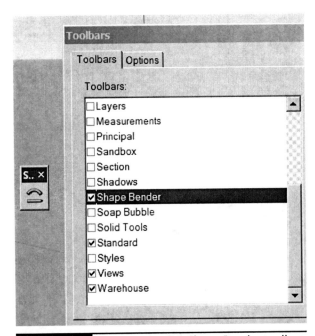

Figure 4-19 Turn on the CTL Shape Bender toolbar.

4. Highlight the chair and then click on the CTL Shape Bender tool. The cursor will turn into an arrow and line (Figure 4-20).

5. Run the cursor over the red axis line drawn in step 2. When it turns blue, click it. The words "Start" and "End" will appear (Figure 4-21).

6. Run the cursor over the Bezier curve drawn in step 1, and click. The chair will extrude along its length (Figure 4-22). Hit the UP ARROW key on the keyboard to reverse its position if needed. Then hit RETURN to set.

7. The bench (Figure 4-23) needs some cleaning up, which is common with this extension. The extra face that formed near the curve can be selected and deleted. The thick line along the back of the seat is actually two double lines that can be selected and deleted. The Bezier curve also should be deleted. You might want to copy the legs, add a few more, and perhaps extrude them around the Bezier curve, too.

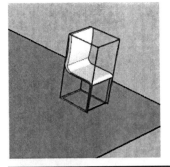 

Figure 4-20 Hightlight the chair and then click on the CTL Shape Bender tool.

 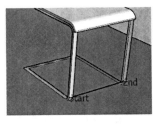

Figure 4-21 Click the tool onto the red axis line to select it. "Start" and "finish" will appear.

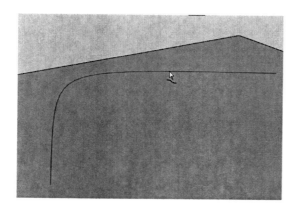

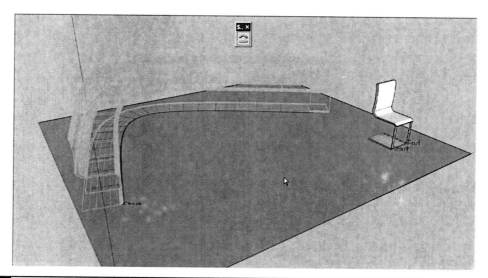

Figure 4-22 Click the Bezier curve to extrude the chair. Hit the UP ARROW key to reverse its position if needed.

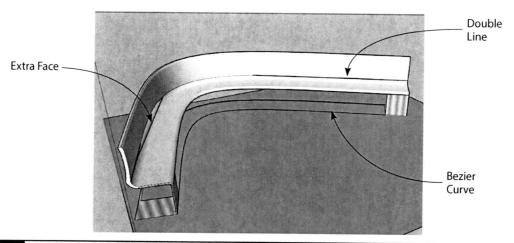

Figure 4-23 The extruded bench.

Fake Fingernail Project

How about some fake fingernails for Halloween? Let's make one (Figure 4-24) with the CTL Shape Bender and Shell. Take a straight-down photo of your index finger to use for tracing. This will make the nail truly custom to your finger.

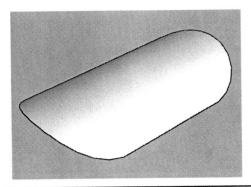

Figure 4-24   Fake fingernail.

1. Click File > Import, make sure the Use as Image button is checked, and import the photo. Don't worry about its size because we'll scale it later. Click the Top view on the Views toolbar. Then activate the Two Point Arc tool, and draw an arc over one cuticle

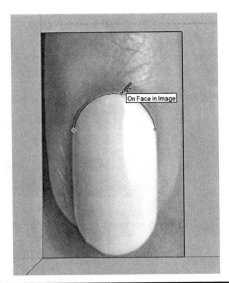

Figure 4-25   Import a photo of a finger and draw an arc on one cuticle.

(Figure 4-25). Remember to look for the On Face inference with each click, and to pull the arc straight along the green axis. Just as with a circle, you can change the arc's default number of segments by typing the number you want immediately after clicking it onto the work plane.

2. Copy and mirror the arc (Figure 4-26). Use MOVE + CTRL (or CTRL + C and CTRL + V) to

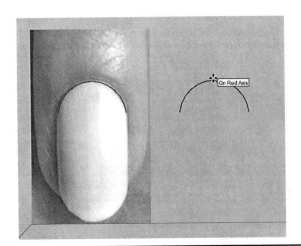
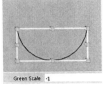

Figure 4-26   Copy the arc, and then mirror it with the Scale tool.

copy it. In Chapter 3 we used Flip Along in the context menu for mirroring, but that function doesn't show up on a non-grouped arc. Instead, click the Scale tool on the arc, push the top-middle grip down a bit, and then type **–1**. The arc will flip.

3. Draw a straight line down from the first arc, and connect it to an endpoint on second arc (Figure 4-27). If the line is too long, move the arc up the line's length, and trim excess with the Eraser.

4. Draw the nail's other side to finish the shape (Figure 4-28). A face will form. If the blue side is up instead of the white side, select the whole nail, right-click, and choose Reverse Faces.

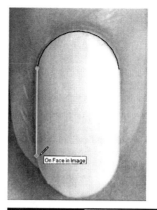 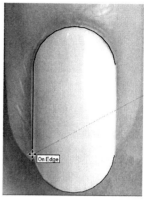

Figure 4-27 Draw a line down from the first arc, and connect the second arc to it.

5. **OPTIONAL:** If you want to extend the nail's length, select an arc, click the Move tool on it, and stretch it along the green axis

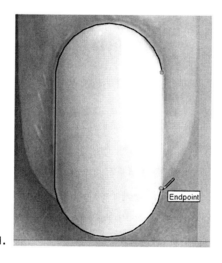

1.

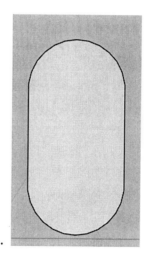

2.

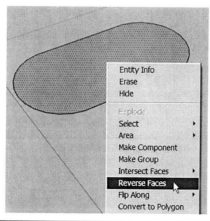

3.

Figure 4-28 Finish the nail's shape, and reverse faces if needed, so that the white face is on top.

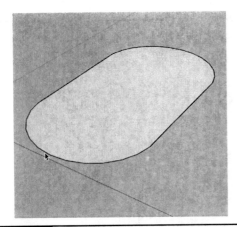
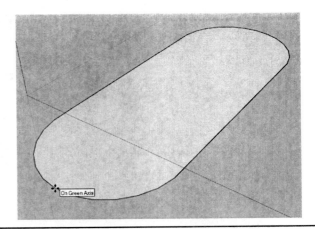

Figure 4-29   To make the nail longer, select an arc, and stretch it with the Move tool.

(Figure 4-29). If you want to customize the top of the nail with any kind of geometry, now is the time to do it.

6. Group the nail. Then make a rectangle to use as a drawing surface and group it. Move the nail onto the rectangle (Figure 4-30). By now you may have noticed the "flashing" that occurs when two planes are adjacent. This happens because SketchUp doesn't

know which plane to display. It doesn't affect anything, though.

7. Draw a line and an arc to use with the CTL Shape Bender tool (Figure 4-31). Remember to draw the line on the nail along the red axis. Inference the arc from the edges of the nail to make it the same length as the nail. Draw the arc's bulge along the blue axis.

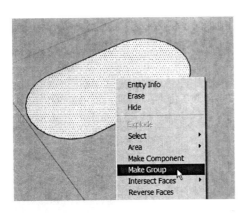
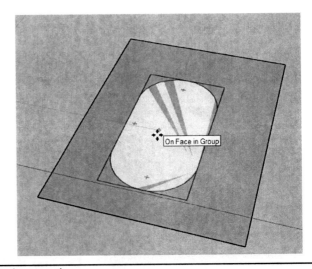

Figure 4-30   Move the grouped nail onto a grouped rectangle.

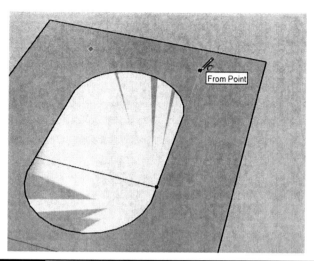
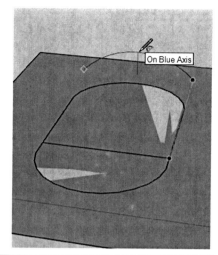

Figure 4-31 Draw a line and arc to use with the CTL Shape Bender tool.

8. Use the CTL Shape Bender extension to curve the nail (Figure 4-32). To recap: select the grouped nail, click on the tool, click on the red axis line, click on the arc, and press RETURN.

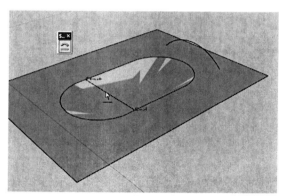

1. Click the Shape Bender tool onto the line.

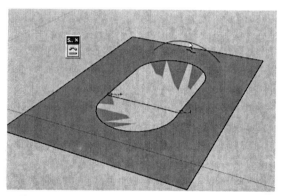

2. Click on the arc.

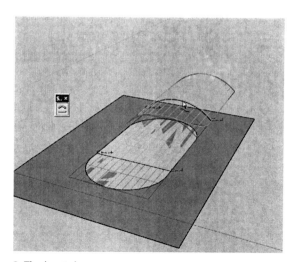

3. The bent shape appears.

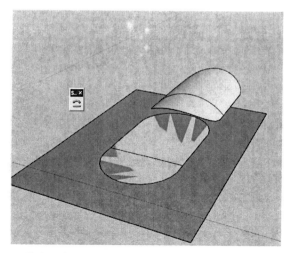

4. Click to finish.

Figure 4-32 Curve the nail with the CTL Shape Bender extension.

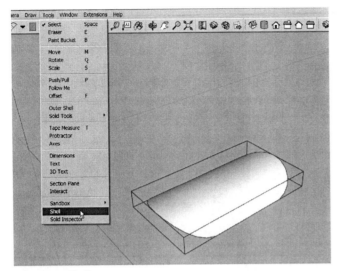

1. Activate Shell.

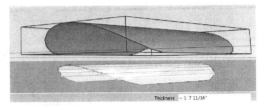

2. A copy will appear.

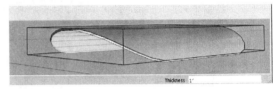

3. Adjust the offset.

Figure 4-33   Use the Shell extension to give the nail thickness.

9. Give thickness to the nail (Figure 4-33). Push/Pull doesn't work on curved surfaces, so use the Shell extension. It offsets groups and components with a second group or component. First, highlight the grouped nail and then click on Tools > Shell. A second component will appear, and you'll see its default offset in the Measurements box. Type the offset you want; because we never scaled the nail, eyeball something proportionate. Then hit ENTER.

10. Scale the nail (Figure 4-34). Measure the width of your fingertip, click the Tape Measure on opposite edges of the model's width, type the measurement, and press ENTER. The nail will scale to that size, customized to your finger! You might need to click on Zoom Extents to find it.

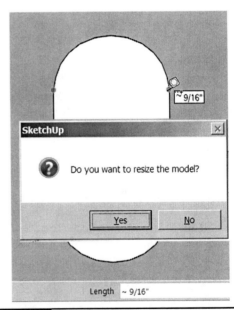

Figure 4-34   Scale the nail to make it 1" long.

Right now the nail consists of two groups: the original and the one made with Shell. Make a copy of this model. Keep the original as-is. On the copy, explode both groups (Figure 4-35). Then select the whole nail and group it. This may help make the model more 3D-printable.

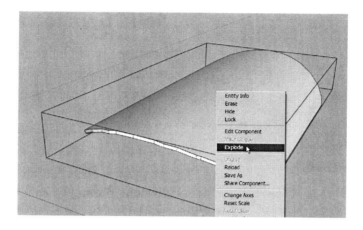

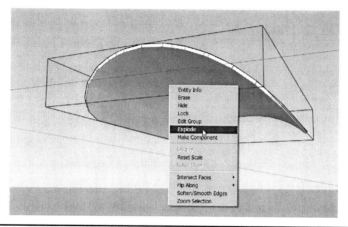

Figure 4-35 Explode the original and offset components, and then make one group that contains both of them.

Done! There's a company that 3D prints nails like this with precious metals. Figure 4-36 shows some of their offerings.

Before moving on to our next project, let's talk about two subjects that will be used in it: Thingiverse, solid geometry, and solid tools.

Figure 4-36 3D-printed nails by https://www.facebook.com/CLAWZco?hc_location=ufi.

Thingiverse

Thingiverse is a model repository. Like the 3D Warehouse, anyone can upload to it, and you'll find thousands of models ("Things") made by professionals and casual users. Unlike the models in the 3D Warehouse, most models in the Thingiverse are covered by a Creative Commons copyright. Thingiverse allows sharing but also maintains owners' rights.

Most Thingiverse files are in .stl format. Sometimes you find other files as well, such as the original model, which enables you to easily edit it if you have the software program it was

Figure 4-37 This Thing consists of three Space Invaders .stl files and two SketchUp files.

made in. Find SketchUp files for Things by using *SketchUp* as a search term. When you find something you like, click on it and then click the Download This Thing! button. Then download and unzip any zipped files. I searched for *Space Invaders* and found the files shown in Figure 4-37. Coincidentally, there were even two SketchUp files.

Solid Geometry

Because SketchUp is a mesh modeler, it makes hollow models. They're just faces and edges, empty inside: think air-filled balloon. Solid modeling programs make models with a continuous volume: think rock. So a SketchUp solid isn't like a solid made with a solid modeling program such as Autodesk Inventor. "Solid" in SketchUp simply means closed geometry. If the model held water, none would leak out.

When you draw a rectangle, push/pull it up and group it, you've made a solid. Loose geometry cannot be a solid; it must be a group or component. Verify whether a group or component is solid by highlighting it, right-clicking, and choosing Entity Info. If it says Solid Group or Solid Component, it's solid. If it just says Group or Component, it's not solid; there's a hole or extra piece of geometry somewhere in it. Figure 4-38 shows examples of solids. Note that the solid's volume is also shown, which is useful when calculating material costs for 3D printing.

Know that it can be tricky to make a solid that's more complicated than a pushed/pulled rectangle. Because a tiny face or edge embedded inside will prevent it from being solid, a solid

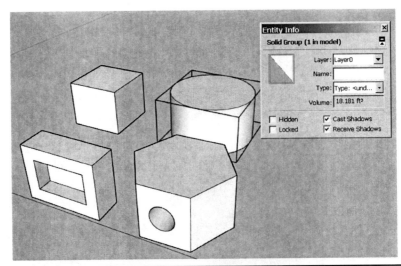

Figure 4-38 SketchUp solids are closed geometry made into a group or component. The Entity Info box verifies if a model is solid.

group often becomes unsolid very quickly. We'll discuss troubleshooting tools in Chapter 5.

Solid Tools

Solid tools let you do things that would be impossible or take a lot of steps with native tools. SketchUp has six; activate them at Views > Toolbar > Solid Tools. The first one on the toolbar, Outer Shell, is available with Make. The rest are Pro features. Solid tools only work on solid geometry, and they work differently than native tools. Basically, you push solids together and then choose what operation you want to perform by picking a specific Solid tool. Once a Solid tool operation has been performed, a model's parts can't be individually edited anymore. So it's good practice to save a copy of the model first.

In the following project we'll use the Subtract tool to subtract one solid from another.

Chocolate Mold Tray Project

Let's make a tray with custom molds to fill with chocolate, Jello, or ice cubes (Figure 4-39).

Install the SketchUp Stl extension. This enables the import of .stl files, which are basically shells around mesh models. Then go

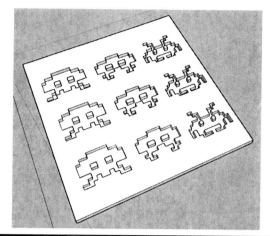

Figure 4-39 A chocolate mold tray made with a downloaded .stl file and Solid tool.

to File > Import and scroll through the Files of Type field to Stereo Lithography Files (Figure 4-40). Navigate to the one you want and click Open (or hit ENTER). The .stl files display the icon of a program that can open them, in this case 123D Meshmixer. Click the Options button, check Merge Coplanar Faces, set the units to the receiving file's units, and click Preserve Drawing Origin. The file will import at SketchUp's origin and may be very small; click Zoom Extents to find it (Figure 4-41).

I imported the three Space Invaders .stl files downloaded from Thingiverse. Here are steps to make the tray:

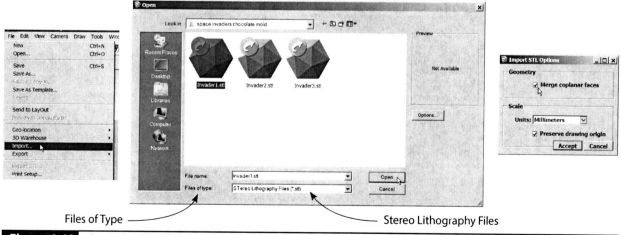

Files of Type ———   ——— Stereo Lithography Files

Figure 4-40 Once the SketchUp Stl extension is installed, you can import an .stl file.

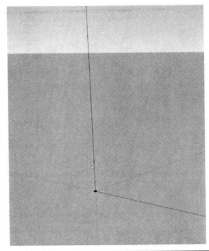

Figure 4-41   The .stl file will import at the origin. You may need to click Zoom Extents to find it.

1. Group each model separately; then rotate and copy them (Figure 4-42).

2. Draw and group a rectangle, and move it under the models (Figure 4-43). Adjust the rectangle's size as needed by highlighting and stretching its edges with the Move tool. Then give it some thickness with Push/Pull (remember to open the group editing box) to make it a base.

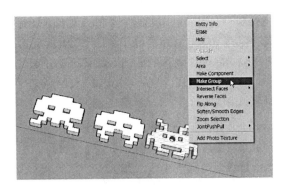
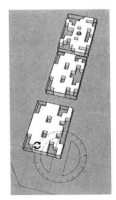

1. Group each model.

2. Select and rotate them.

3. Copy them.

Figure 4-42   Group the models, rotate for better positioning, and copy.

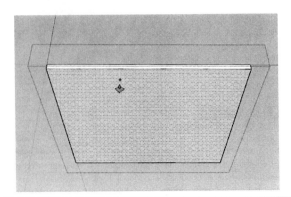

Figure 4-43   Place a grouped rectangle under the models and give it thickness.

3. Move the models part way into the base (Figure 4-44). Not all the way, just part way. The amount they're pushed inside the base determines the mold's depth. Pushing the models almost all the way down will make deep molds, and pushing them in just a little will make shallow molds. Press and hold the

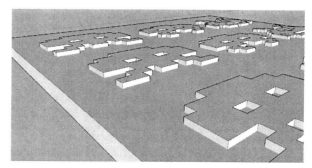

Figure 4-45 Edges did not form when the models were moved into the base.

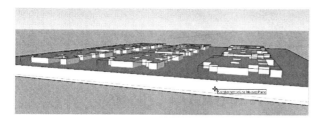

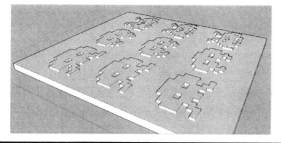

Figure 4-44 Move the models partway into the base.

UP ARROW key to inference lock them along the blue axis, which makes moving them straight down easier. If they won't move down at all, right-click on them and choose Unglue.

Note that there is no edge line between the Space Invaders and the base (Figure 4-45). We've used the Intersect Faces with Model function before to create one, but now we're going to use the Solid Subtract tool.

4. On the Solid Modeling toolbar, activate the Subtract tool. Click a Space Invaders figure first and the base second (Figure 4-46). The

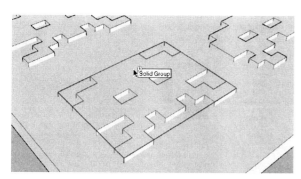

1. Click on the figure.

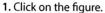

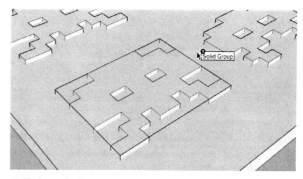

2. Click on the base.

Figure 4-46 Subtract the Space Invaders model from the base with the Subtract tool.

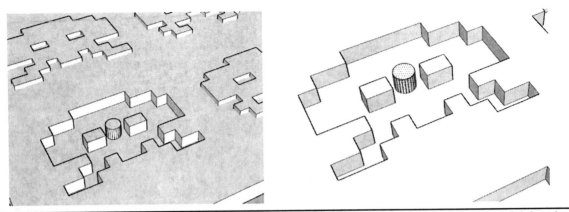

Figure 4-47 After this figure is subtracted from the base, the exposed cylinder can be selected and deleted.

Space Invaders model will get subtracted from the base. Note the exposed cylinder; select and delete it (Figure 4-47).

The order in which you click makes a difference because the first item gets subtracted from the second. When finished, scale the whole model by clicking the Tape Measure on opposite corners and typing the dimension wanted. Done!

Architectural Terrain Model Project

Terrain models show a patch of ground's topography: slope of the earth, vegetation, and built features. They're used to study a site and building placement. Ours uses an extension that isn't in the EW. Point your browser to http:// sketchucation.com/. Make an account, log in, and click on the Plug-in Store tab (Figure 4-48).

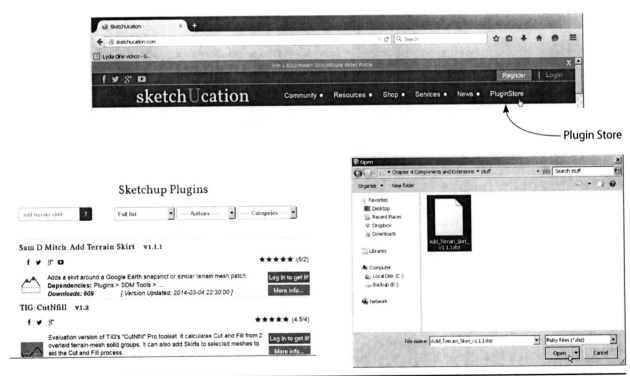

Figure 4-48 Download the Add Terrain Skirt extension from sketchucation.com and install it at Window > Preferences > Extensions.

Then search for *Add Terrain Skirt*. Download the file, and install it by clicking on Window > Preferences > Extensions and on the Install Extension button.

1. Find the general area (Figure 4-49). Click on File > Geo-location > Add Location. This accesses a world map. Type what you're looking for in the search field—an exact address, name of city, state, even

landmark. The slider on the left lets you zoom in and out.

2. Find the exact area (Figure 4-50). Once you've zoomed in to the general area, click on Select Region. Four pins will appear. Move them to select the exact patch wanted for the model. Include a feature whose size you know or can reasonably guess. Then click Grab.

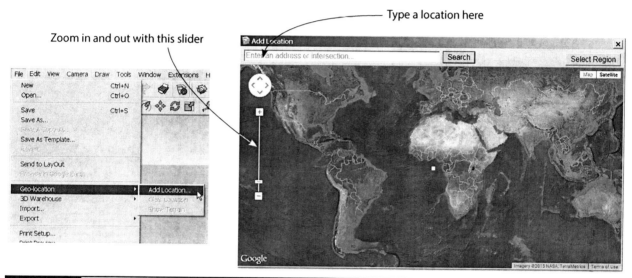

Figure 4-49 Search with the Geo-location feature.

Figure 4-50 Select the exact area for the model by adjusting the pins.

3. Unlock the patch (Figure 4-51). The patch of Google Earth terrain will load into the model. It's a grouped photo. Click on it to select. Note that the group border is red, meaning that it's locked. Right-click and choose Unlock. It should enter to scale, but if it doesn't, click the Tape Measure on opposite ends of a feature whose size you know, and type the correct size. The whole model will scale proportionately.

4. Access the patch's slope (Figure 4-52). Google Earth terrain comes with a turned-off *layer* or sheet containing slope information. Click on Window > Layers and check the box next to Google Earth Snapshot. This makes that layer visible.

Figure 4-51 Google Earth layers.

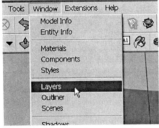

1. Click open the layers window.

2. Make the Google Earth Terrain visible.

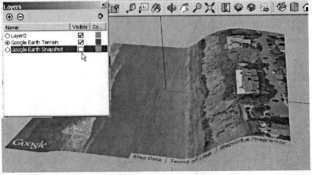

3. Make the Google Earth snapshot layer invisible.

Figure 4-52 Google Earth layers.

Then uncheck the Google Earth terrain patch because we don't need it anymore.

5. Add volume to the model. Clicking on the slope, you'll see that it's another locked group. You can right-click to unlock it if you want, but it's not necessary. Select it, then click on Extensions > Add Terrain Skirt (remember, we installed it earlier). Volume will automatically be added to the selected group (Figure 4-53), and you can push/pull it thinner or thicker. Done!

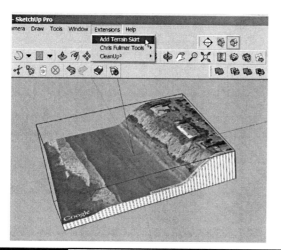

Figure 4-53 The Add Terrain Skirt extension adds thickness to the terrain model.

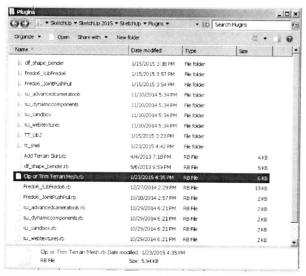

Figure 4-54 The Plug-ins folder on a PC, located in the program files, shows all installed extensions.

If you want to crop the watermark on the Google Earth Terrain picture or trim overlapping mesh from multiple patches, use the Clip or Trim Terrain Mesh extension from sketchucation.com. Install this extension by downloading and dragging it manually into the plug-ins folder in your computer's program files (the path was described earlier in this chapter). Figure 4-54 shows the plug-ins folder on a PC. All extensions live here. If an extension becomes corrupted, delete it from this folder and reinstall it. To completely remove it, delete it here.

Pacman Ghost Ornament Project

This project uses the Outer Shell and Intersection solid tools and the s4u Face Maker extension to make the Pacman ghost ornament in Figure 4-55. We'll draw it large and scale it down when finished.

1. Model and group a block to draw on (Figure 4-56). Click the Rectangle tool on the workspace, type **8',8'**, and then push/pull it up 16'. Group it and click the Front View icon.

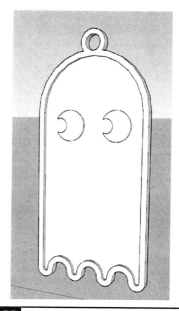

Figure 4-55 A Pacman ghost ornament.

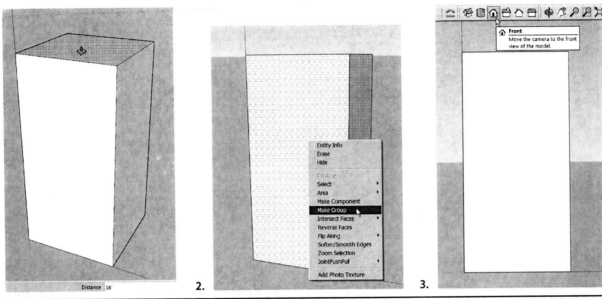

Figure 4-56 Model and group a block.

2. Draw the ruffle (Figures 4-57 and 4-58). Click on the Two Point Arc tool, click the endpoints 1' apart, and then pull the bulge down until the Half Circle inference appears. Click to place. Make a second arc the same size in the opposite direction. Group them together.

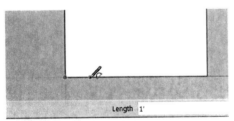

1. Click the arc's endpoints.

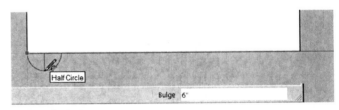

2. Drag the bulge.

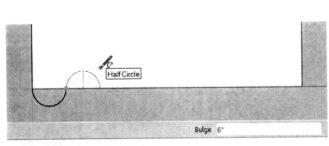

3. Make an identical arc opposite it.

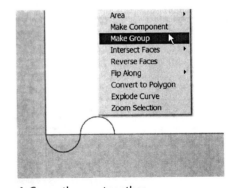

4. Group the arcs together.

Figure 4-57 Draw the first two arcs of the ruffle, and group them together.

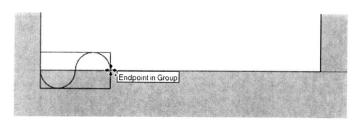

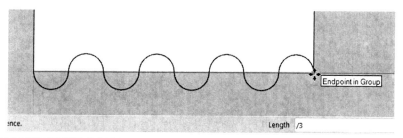

Figure 4-58 Make multiple, equally spaced copies of the grouped arcs.

Click on Move and then press the CTRL key to signify multiple copies (plus and minus signs will appear near the cursor). Drag a copy to the opposite end, and click. Then immediately type *l3*. Two more copies will appear, equally spaced between the first and last.

3. Make the ghost's sides (Figure 4-59). First make the ruffle symmetrical by deleting the extra arc on the right (open its group editing box to do this). Then draw vertical lines from the arcs' endpoints to the top of the block.

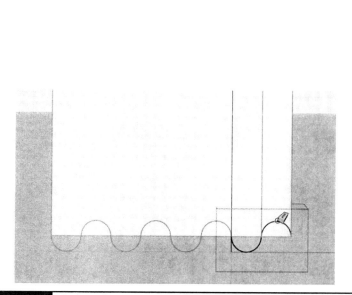

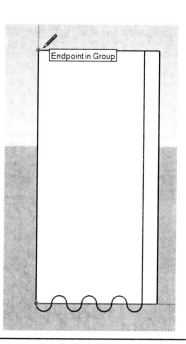

Figure 4-59 Delete the extra arc and draw the ghost's sides.

4. Draw the ghost's head (Figure 4-60). Use the Two Point Arc to draw a half-circle, and then trim the sides as needed. You can use the Eraser, but SketchUp 2015 lets you trim it with the Two Point Arc, too. Double-click on the extra part of the line as soon as you finish drawing the arc, right-click, and choose Erase. Then erase the block behind it, and explode the ruffle groups at the bottom (Figure 4-61). You'll be left with an outline. If there are some lines and planes inside it, erase them, too.

5. Fill the outline with a face (Figure 4-62). Turn on the s4u Make Face tool by clicking on it at Views > Toolbars. Then select the outline, and click the tool. If the outline is coplanar (as it should be because you drew it all on the block and watched for On Face extensions, right?), a face will form.

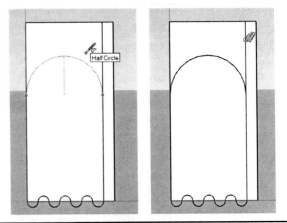

Figure 4-60   Draw the ghost's head.

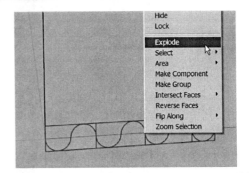

Figure 4-61   Explode the ruffle groups.

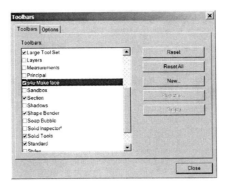

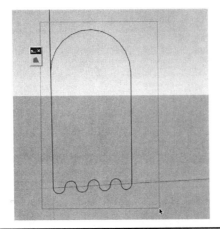

Figure 4-62   Fill the outline with the s4u Make Face extension.

6. Give the face volume (Figure 4-63). Push/pull it forward 4'. Click the Offset tool on the perimeter, and drag a 3' border. Push/pull the border 2' forward. Then group it. Make sure that its Entity Info box lists it as a Solid Group.

7. Model the eyes (Figure 4-64). Draw a circle, push/pull it forward, and group it (verify that it's a solid group through the Entity Info box). Cut a moon shape into it as shown in Figure 4-64. Again, verify that it's a solid group. Then copy it with Move and Ctrl (Figure 4-65).

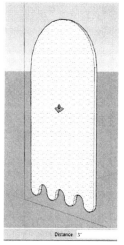

1. Push/pull the face.

2. Offset the perimeter to make a border.

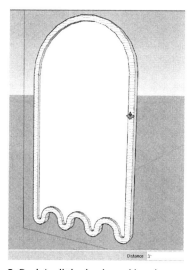

3. Push/pull the body and border.

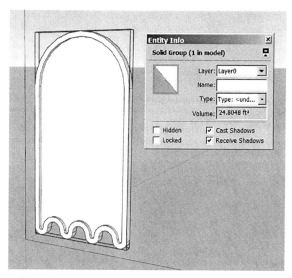

4. Group it.

Figure 4-63 Create a ghost from the outline.

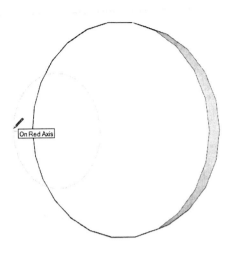

1. Place a small circle on a larger one.

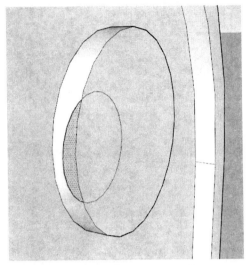

2. Delete the left part of the small circle.

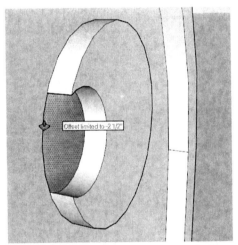

3. Push/pull the remainder of the small circle back.

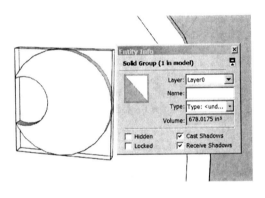

4. Group it.

Figure 4-64 Model an eye.

Figure 4-65 Copy the eye.

8. Weld the body and eyes together (Figure 4-66). Right now they're three separate groups and have to be selected together to move together. We'll use the Outer Shell tool to weld them together. Click on it, and then click on the body and one eye. Then click on the body and the other eye. All three should now highlight when you click any one of them and can be moved together. Welding them together is essential to making the ornament 3D-printable.

9. Model a top ring (Figure 4-67). Draw two concentric circles, delete the center, move them to the ornament's front edge, and push/pull them. Click the Push/Pull tool onto the ornament's back edge to make the ring precisely the ornament's width. Group it and verify that it's a solid group. Then push the ring down into the body. Hold the keyboard's DOWN ARROW to lock the ring along the blue axis if you're having trouble moving it straight down.

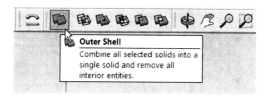

1. Click on the body.

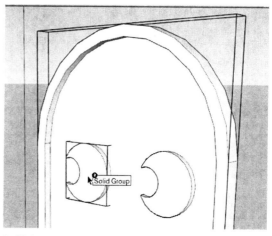

2. Click on one eye.

Figure 4-66 Use the Outer Shell tool to weld the three groups together.

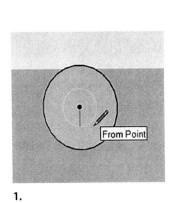

1.

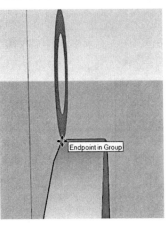

2.

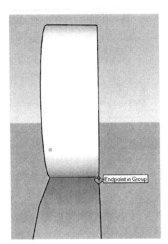

3.

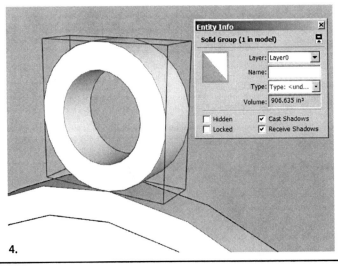

4.

Figure 4-67 Model a top ring.

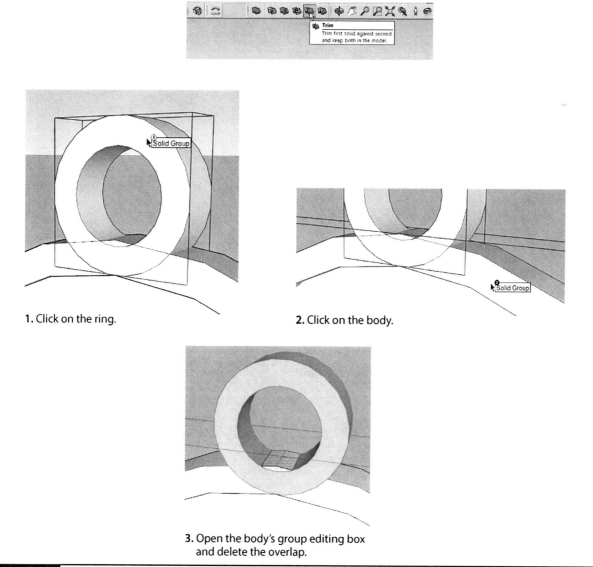

1. Click on the ring.

2. Click on the body.

3. Open the body's group editing box
and delete the overlap.

Figure 4-68 Trim the overlap inside the ring with the Trim solid tool.

10. Delete the overlap inside the ring (Figure 4-68). Click on the Trim tool, click on the ring, and then click on the body. Next, click open the body's group editing box and delete the overlap inside the ring.

11. Weld the ring and the body together (Figure 4-69). Click on Outer Shell, click on the ring and then click on the body. When finished, the ornament should highlight as one solid piece when you click anywhere on it.

Finally, scale the ornament by clicking the Tape Measure onto opposite ends and typing whatever size you want.

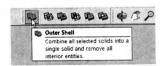

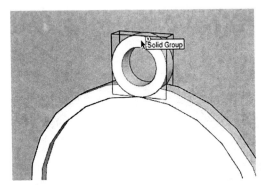

1. Click on the ring.

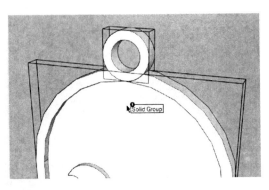

2. Click on the body.

3. The combined ring and body.

Figure 4-69 Weld the ring and body together with the Outer Shell tool.

AutoCAD Floor Plan Project

This is for you AutoCAD users! Pro can import a .dwg file. It imports as SketchUp geometry. In this project we'll import and model the floor plan in Figure 4-70.

1. Import the AutoCAD file (Figure 4-71). Click File > Import, and locate it. Make sure that AutoCAD Files is visible in the Files of Type field at the bottom. Then click Options, and check the two upper boxes. The top two boxes tell SketchUp to

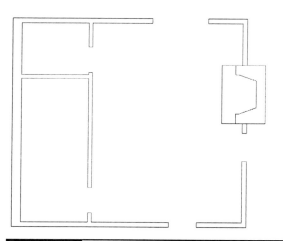 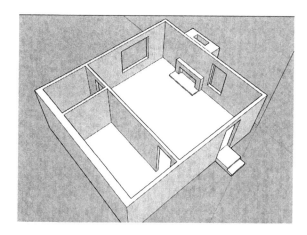

Figure 4-70 An AutoCAD floor plan and the SketchUp model made from it.

AutoCAD software doesn't need to be installed on your computer, but a .dwg file needs some preparation before import to make it easy to model. This is optimally done within AutoCAD. When importing your own files you may want to:

- Copy and paste the plan into a new AutoCAD file to prevent stale metadata importing into SketchUp. Then move the plan to the origin.

- Run the Purge and Audit commands to clean up any old data that did enter the new file.

- Ensure that lines that are supposed to be connected at endpoints are indeed connected.

- Run the Units command so that you know what the units are. Scale the file to 1:1 if it isn't already scaled.

- Delete unused layers. SketchUp automatically discards anything in the imported .dwg file that has no 3D relevance, such as text, dimensions, hatch lines, and title blocks. However, it doesn't discard the layers they're on.

- Explode all polylines, arcs, and filleted lines.

- Erase any construction entities, such as points created where lines were divided. They import as random bits of geometry.

- Remove all textures, x-referenced and imported files, colors, and dynamic blocks.

- Ensure that the file is smaller than 15 MB because larger ones might not import.

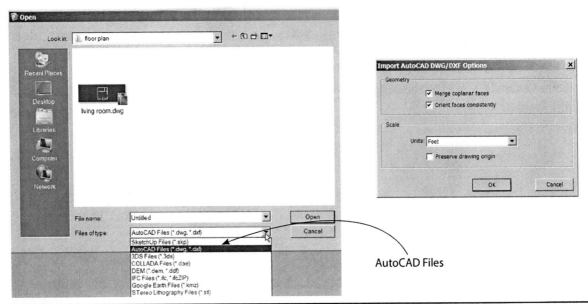

Figure 4-71 Navigate to the AutoCAD file and check the appropriate boxes.

treat the AutoCAD file like a SketchUp file. Set the Units field to the AutoCAD plan's units. For example, if the plan's units are feet, set it to feet. This is important! If the AutoCAD plan's scale and this box's scale are different, scaling the plan after import will be difficult. If you uncheck the Preserve Drawing Origin box, SketchUp can place the imported AutoCAD plan at the origin. If it imports far from the origin, clipping may occur, a glitch that causes part of the plan to disappear.

Click Open to import the .dwg file. A box will briefly appear that lists the specific data

imported, and then the AutoCAD file will appear. It's now SketchUp geometry.

> If your plan imports as a group, explode it (select, right-click, choose Explode). One explosion shouldn't affect imported blocks, which enter SketchUp as components. However, you might find that some blocks (now components) behave oddly when panning and orbiting. Fix by exploding and remaking them as components again.

2. Make faces (Figure 4-72). The plan imports as lines, with no faces between them. Drag

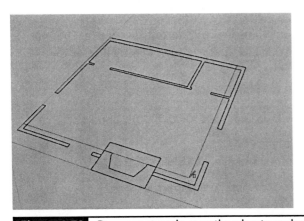 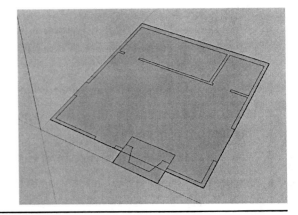

Figure 4-72 Drag a rectangle over the plan to make faces.

the Rectangle tool over it (click on opposite corners) to cover the whole plan with a face. The existing lines will break it into multiple faces.

3. Scale the plan (Figure 4-73). Click the Tape Measure on opposite ends of an item with a known dimension. Here I've clicked on a door opening. After clicking on the second endpoint, immediately type the desired length. A note will appear asking if you want to rescale the model. Click Yes, and the model will scale proportionately to the size of the item you scaled. Note that the plan here has the front (white) side down, so select, right-click, and choose Reverse Faces to face it up.

4. Model the walls and floors (Figure 4-74). Push/pull one of the wall sections up,

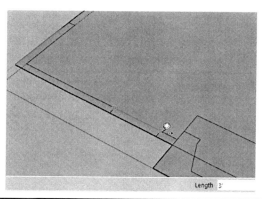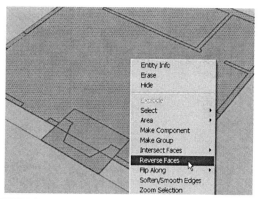

Figure 4-73 Scale the model by scaling an item with a known dimension. Reverse faces, if needed, so that the front (white) side shows up.

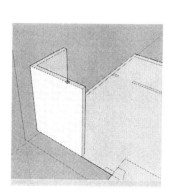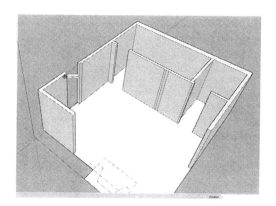

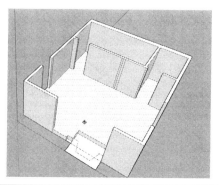

Figure 4-74 Model the walls by pushing/pulling the first section up and inference matching the rest to it. Then do the same to the floors.

type **9'**, and press ENTER. Then inference match the rest of the walls to that section by pushing/pulling them up, hovering the cursor over the first section, and releasing it. Push/pull up the floor in one room 6" and inference-match the other floors to it.

5. Finish the window and door openings. When you pull up walls with windows marked in them, they'll be unfinished at the top (Figure 4-75). Draw lines across the top, mark the head height with the Tape Measure, and trace over the guideline created (Figure 4-76). Do this on both sides of the wall.

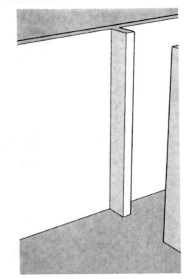

Figure 4-75   After the walls are push/pulled up, openings will be unfinished at their tops.

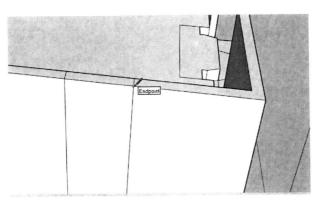

1. Draw lines across the top.

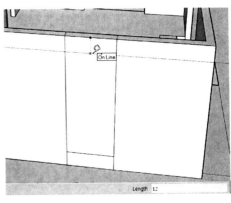

2. Draw a guideline 12" from the top.

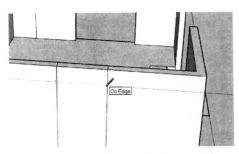

3. Trace over the guideline with the Pencil.

Figure 4-76   Draw the top of openings with the Tape Measure and Pencil tools. Do this on both sides of the wall.

Faces will form when you draw new lines; just delete them (Figure 4-77). Eventually the openings should look like the ones in Figure 4-78. Note that there must be faces on the sides, too. If a face got deleted, undo and trace your steps to see what went

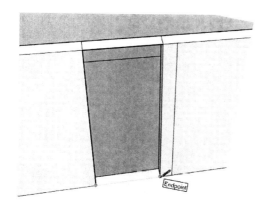

1. Draw a line.

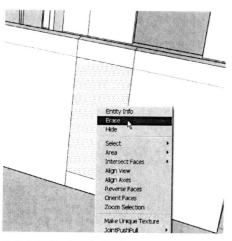

2. A face forms.

3. Delete the face in the front.

4. Delete the face in the back.

Figure 4-77   Delete the faces on both sides of the wall.

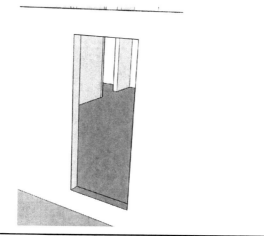

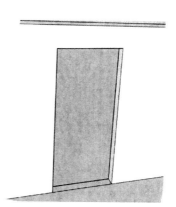

Figure 4-78   A finished opening, front and back.

wrong, and try another workflow (sequence of steps). Inside the house, delete all unnecessary lines (Figure 4-79).

6. Draw stairs. Because we push/pulled the floor up, the inside isn't level with the outside. Fix that with stairs as shown in Figure 4-80.

7. Model the fireplace (Figure 4-81). Do this by pushing/pulling its outline up. Or import and scale one from the Warehouse.

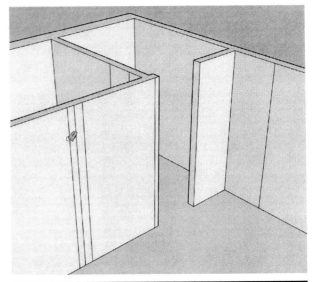

Figure 4-79 Delete all unnecessary lines inside the house.

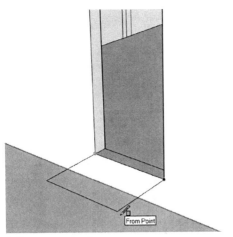

1. Draw a rectangle.

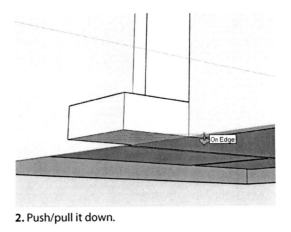

2. Push/pull it down.

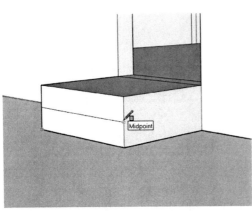

3. Draw a line at the midpoint.

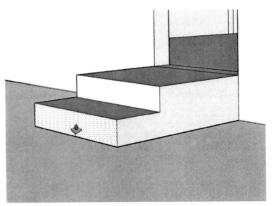

4. Push/pull it forward.

Figure 4-80 Model some stairs in front of the door.

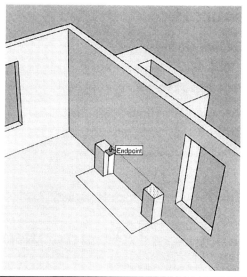

Figure 4-81 Model the fireplace. Here one side is being inference matched to the other.

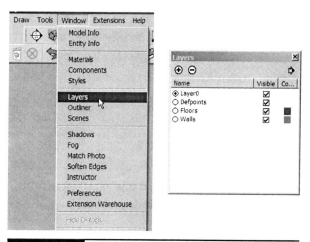

Figure 4-82 At Window > Layers, click the plus sign to make and name new layers.

Layers

Layers are sheets attached to a model. You can move groups and components onto these sheets. All models start with Layer 0, but you can add as many as you want. Go to Window > Layers, and click the plus sign (Figure 4-82) to create and name a new layer. Then select geometry,

make it a group or component (Figure 4-83), right-click it to access the Entity Info box, and click on the box's drop-down arrow to select a specific layer. This puts the selected group or component on that layer. You can turn layers off, which is useful when working on parts obscured by other parts. To delete a layer, right-click and choose Delete. Layer 0 cannot be deleted.

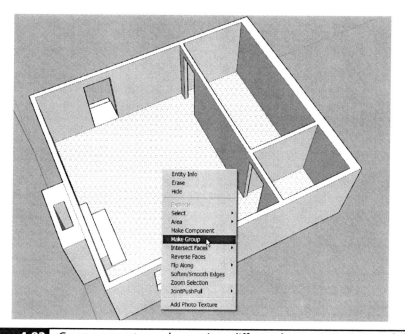

Figure 4-83 Group geometry and move it to different layers via its Entity Info box.

All loose geometry should be on Layer 0. Don't put loose geometry on any other layer because it will be easy to lose. Make a layer current, meaning that all new construction will go on it, by clicking its radio button.

Summary

In this chapter we installed and used extensions, which are tools that perform actions the native ones can't. We included content from the Thingiverse in our modeling, discussed solid geometry, used solid tools, and imported an AutoCAD file. In Chapter 5 we'll make some of our models 3D-printable.

Resources

- **Creative Commons:**
  http://creativecommons.org/
  Information on public copyright licenses.

- **Extension Warehouse:**
  https://extensions.sketchup.com/
  Direct Web URL

- **Terrain Model Video:**
  https://youtu.be/y22kYRnsuus

Repositories of Free and Pay Models

- **Thingiverse:** http://www.thingiverse.com/

- **YouMagine:** https://www.youmagine.com/

- **GrabCad:** https://grabcad.com/

- **Formfonts:** http://www.formfonts.com/

- **Pinshape:** https://pinshape.com/

Student Pro Licenses

- http://www.creationengine.com/html/
  m.lasso?m=GG&gclid=CJ7h9eOO98M
  CFYQ8aQodcZcAag

CHAPTER 5

Making the Model 3D-Printable

So far our models have only lived in the computer. To become physical reality they must meet certain requirements. In this chapter we'll discuss what those requirements are and how to meet them.

Design Considerations

While modeling, you are constantly deciding which features to show and how to show them. When printing that model on your own machine, you have control over the result. However, when you use an online service, assumptions may be made during the automated process resulting in a product that isn't what you intended. In fact, this can happen on your home machine, too. So always be cognizant of a design's suitability for printing while modeling it to avoid having unprintable models "fixed" into ones that don't resemble what you wanted.

A printable model:

- **Has thickness** (Figure 5-1). Zero-thickness planes are meaningful to SketchUp but not to a 3D printer. For instance, a single-face wall and the scale figure are not printable.

- **Has appropriate thicknesses.** Models are often scaled down before printing, but a part that is scaled down too much won't print. Features that start thin or small get thinner or smaller. Thin-walled models may break

during printing or shipping or not print at all. Thus they must be arbitrarily thickened or deleted. This also applies to models that have a large mass connected to a thin mass. Minimum thickness varies with the chosen material, but to be safe, use 2 mm as a minimum size when printing with plastic. It's best to scale the model to its final printed size within SketchUp even if it is possible to do so in the slicer (3D printing) software.

- **Considers plastic shrinkage.** ABS (oil-based) filament shrinks about 2 percent. PLA (plant-based) filament shrinks about 0.2 percent, which makes it a better choice when dimensional accuracy is important.

- **Leaves enough clearance (space) between moving or separate parts such as links,**

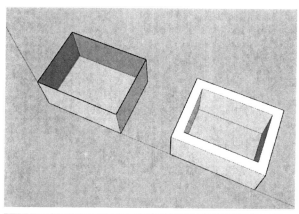

Figure 5-1 The left model is single-face and unprintable. The right model has thickness, which is needed for printing.

gears, and cogs (Figure 5-2). Insufficient clearance results in a welded-together print. A clearance of 0.4 to 0.5 mm on all sides usually works. Parts that must fit snugly together need between 0.1 and 0.25 mm of

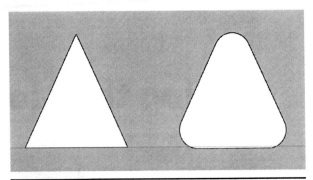

Figure 5-3   Round off sharp features.

clearance on all sides to ensure they don't easily come apart.

- **Has rounded, not pointed, features** (Figure 5-3). The round printer nozzle can't make sharp corners.

Structural Considerations

Appropriate thicknesses and clearances alone don't make a model printable. The digital file also must be "watertight," meaning that if you poured water into it, none would leak out. Holes may cause the slicer software to reject the file, fix it with unwanted results, or cause the printer to stop when it reaches the defective part. Hence a model must meet these additional requirements:

- **Have no holes, extra faces, or extra edges in its mesh** (Figure 5-4).

Figure 5-2   Leave appropriate space between separate parts.

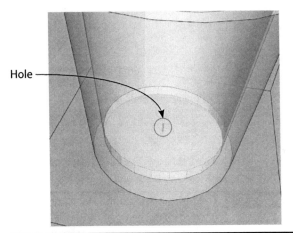

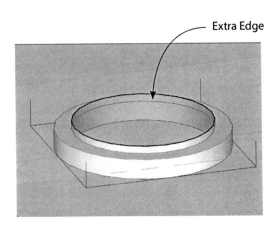

Figure 5-4   The travel mug modeled in Chapter 3. On the left is a pinhole; on the right is an extra edge.

- **Be manifold.** This means that each edge must connect to exactly two faces. Non-manifold edges connect to three or more faces, don't connect to any faces, have overlapping vertices, or have overlaid faces (the last is what causes SketchUp's flashing). This is an issue because each vertex has a unique X, Y, Z coordinate that the printer reads, and overlapping coordinates confuse it. Figure 5-5 shows the many configurations non-manifold edges can take.

- **Have front-facing polygons.** The front (white) side must face out or face up. This is because the slicing software and printer recognize front and back faces and get confused when they aren't oriented properly. Plus, if you have color on the model (discussed later in this chapter), the color must be applied to the front face for the printing software to read it.

- **Be in a format that a 3D printer recognizes.** The most common is an .stl file, but there are others, such as .obj, .vrml, and .3mf files. You cannot send SketchUp's native .skp format—or any program's native format— to a printer. The .stl extension we used in Chapter 4 to import .stl files also lets us export .stl files (Figure 5-6).

- **Be one solid piece with no surface intersections.** Grouping parts together isn't enough; they must be permanently welded in the type of operation that SketchUp's solid tools perform. In Chapter 4 we learned how to tell if a group or component is solid by clicking on its Entity Info box. If it indentifies it as a Solid Group or Solid Component, that part is solid (Figure 5-7). If it only says Group or Component, it's not.

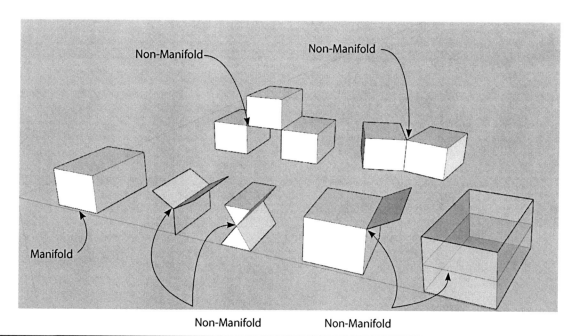

Non-Manifold

Non-Manifold

Manifold

Non-Manifold

Non-Manifold

Figure 5-5 Manifold and non-manifold edges.

As you develop the model, it helps to turn parts of it into solid groups. This also lets you change a feature later without disturbing the rest of it. When all the model's parts are solid, weld them together with SketchUp's Union or Outer Shell solid tools. Union leaves all internal geometry intact; Outer Shell removes it. Use Outer Shell when just the exterior of the model is needed.

Generally, it's not a good idea to explode individual groups before welding them all together because the model might develop problems when reverting back to loose geometry that will keep you from making all of it one solid

Figure 5-6 The .stl extension exports .skp files as .stl files.

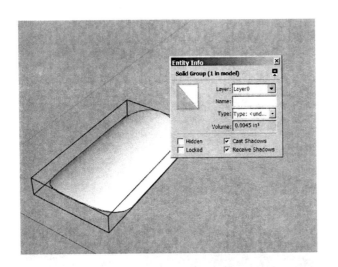

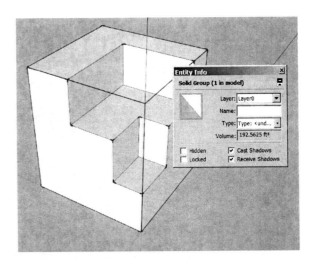

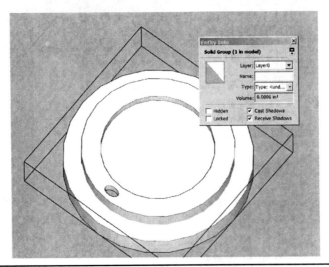

Figure 5-7 The fingernail, icon, and mug lid are all solid groups.

group. We'll see later in this chapter how such a situation can be difficult to fix.

> **Tip:** The Intersect Faces with Model function creates the same intersections as Solid tools but requires manual deletion of extra geometry. Its advantages are that it's on Make, whereas most of the solid tools are only on Pro, and that it works on all models whether they're solid or not.

As you develop a model, keeping a solid group solid becomes increasingly difficult. The workflow (order of steps) can make the difference between the result being solid or non-solid, even when the end result looks the same. Holes and extra geometry often occur after extrusion operations. The tiniest extra piece or reversed face turns a solid into a non-solid, and you'll have to search for and correct it.

Manually Inspect for Defects

If the Entity Info box says that a group or component isn't solid, look for visible defects. To do this:

- Turn on hidden geometry at View > Hidden Geometry.

- Make the model transparent at View > Face Style > X-ray.

- Turn on all layers at Window > Layers to find nested groups and components and loose geometry on layers other than Layer 0.

- Verify face orientation by opening the Entity Info box. At the top are two panels. The left is the front face; the right is the back face. The left should be white/gray. These

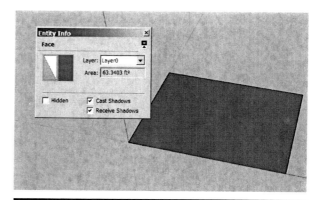

Figure 5-8   The Entity Info box shows that the front side is facing up.

panels are especially helpful when looking at painted surfaces. Figure 5-8 shows that the front is facing up, as it should be. If the red color appears in the left panel, select the face, right-click and choose Reverse Faces.. Then, if appropriate, paint the front face and restore the back face's default color.

- Use the Section tool (Figure 5-9) to inspect inside the model. Turn it on at View > Toolbars > Section, and click the first icon to activate it. A cutting plane will appear and orient to whatever face you run it over. Click it onto the model. Click the Select tool onto the cutting plane to highlight it (it will turn blue). Then click the Move tool onto the plane, and move it slowly through the model, looking for problems (Figure 5-10). Click the Eraser onto the plane to delete it.

Figure 5-9   The first icon is the Section tool. The second toggles the section plane on and off. The third toggles the cut portion of the model on and off.

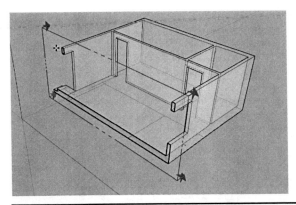 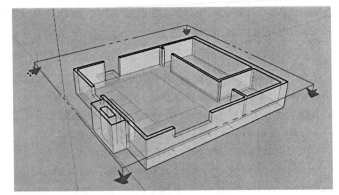

Figure 5-10 Move the Section tool through the model to visually inspect the interior for defects.

The Section tool doesn't actually cut your model; it just displays sectional views of it. You can toggle the cutting plane and the portion of the model in front of it on and off by clicking on the Section toolbar's second and third icons.

Find and delete unwanted faces and edges. Move overlapping geometry. Use the Pencil to recreate missing faces. Fix non-coplanar faces. All this can be tedious, and some defects will likely evade your visual inspection. Others may be difficult to fix manually. Luckily, there are third-party tools to help.

Tools for Making a Model Solid

Anyone who 3D prints with SketchUp needs to be familiar with tools that inspect and fix models for defects that prevent them from being solid. Some popular ones include

- **Solid Inspector². **This is a free extension that is downloadable through the Extension Warehouse. It flags problems in the SketchUp model and fixes simple ones. It works on versions 2014 and newer; there's an older version for earlier SketchUp versions.

- **CleanUp³. **This is a free extension that is also downloadable through the Extension Warehouse. It cleans and optimizes the SketchUp model and helps to keep the file

size down, which is useful whether you plan to 3D print it or not. It also cleans up and simplifies imported .stl files, such as files downloaded from Thingiverse or the 3D Warehouse. This makes them easier to edit. Run CleanUp³ periodically while you model.

- **Edge Tools². **This is a free extension that is downloadable through the Extension Warehouse. It cleans up imported .dxf and .dwg files.

- **SolidSolver. **This is a free extension that is downloadable at www.sketchucation.com. It analyzes groups and components for defects and tries to fix them.

- **SU Solid. **This is an extension available at http://www.susolid.com that inspects and cleans the SketchUp model. There's a limited-feature free version and a full-feature pay version. After paying for the latter (PayPal is an option), the extension will be e-mailed to you. It can only be installed on one computer.

- **Autodesk 123D Meshmixer. **This is a free program available at 123dapp.com that inspects an .stl file for defects. It has automatic and manual repair options. It also lets you mix models together, hollow them out, and weld parts together. Its pay (monthly subscription) option allows commercial use.

- **Netfabb.com** (www.netfabb.com/). You can upload .stl models here for inspection and repair. There are two options: Studio Basic, a free limited-function version, and Studio Professional, a full-feature pay version with manual repair abilities.

- **Microsoft site** (https://netfabb.azurewebsites .net/?hc_location=ufi). This free site is powered by Netfabb. You can upload your .stl and other 3D-printing-format files to it for an automatic repair.

- **Tinkercad.** This is a free solid modeling Web app available at tinkercad.com that is owned by Autodesk. It automatically repairs .stl files on import. Its pay (monthly subscription) option allows commercial use.

- **3D Warehouse Printables.** When you upload a file to the Warehouse, you'll see an "I want this to be 3D printable" option on the upload screen. Check it and your model will be sent to the i.Materialise service bureau for inspection and repair. There's also a checkbox to do this on models you've already uploaded.

Know that all these tools and sites can return different results on the same file, some better than others. A model with a lot of defects may be fixed into something unrecognizable, or may lose parts during an import. It's helpful to know how to use all of them. You might use two or more on one model. We'll use five— Solid Inspector², CleanUp³, SU Solid, 123D Meshmixer, and Tinkercad—to turn some of the models in Chapters 3 and 4 into printable files. Download Solid Inspector² and CleanUp³ from the Extension Warehouse, and install them at Window > Preferences > Extensions (press the Install button). Download the free version of SU Solid from its website, and install it the same way. Download 123D Meshmixer from its website, and follow the installation instructions. Tinkercad is Web-based and works best with the Chrome browser. Both it and 123D Meshmixer require you to set up a free Autodesk account to use.

Solid Inspector²

The Entity Info box of the coffee mug shows that it's not a solid group. Select it and click on Tools > Solid Inspector² (Figure 5-11).

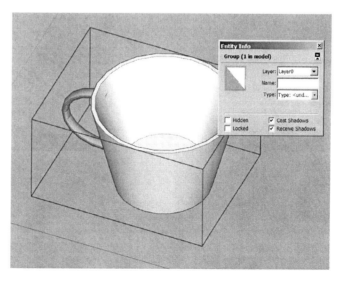

Figure 5-11 Highlight the model, and click on Tools > Solid Inspector².

It runs a quick analysis, highlights problems in red, and returns a window describing them (Figure 5-12). I clicked the Fix All button, and it fixed them. The Entity Info box now shows the mug as a solid group. Well done!

Next, I ran it on the dog tag made in Chapter 3. Again, we see that it highlights and fixes all the problems (Figure 5-13). Results can take several minutes on large, complex models.

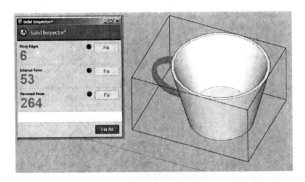

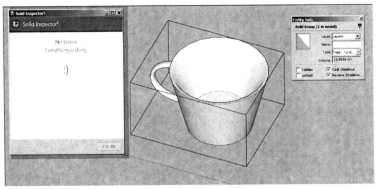

Figure 5-12 Solid Inspector² analyzed and fixed the problems, and now the mug is a solid group.

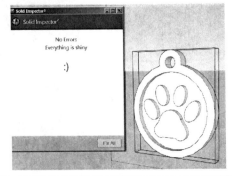

Figure 5-13 Solid Inspector² found and fixed the dog tag's problems.

This tool is very helpful, but it can't fix every problem, and sometimes it returns an "Everything Is Shiny" message when there still are issues keeping the model from being solid. The Entity Info box of the pencil holder showed it as an ordinary group. So I ran Solid Inspector[2] on it. It found flaws, and I clicked the Fix All button. It returned a message saying that it couldn't fix some of the flaws and that I should manually fix them and run the tool again (Figure 5-14).

When this happens, open the group or component's editing box and look for the flaws. You can activate Solid Inspector[2] with the editing box open. I made the model transparent at View > Face Style > X-ray and ran the Section tool through it (Figure 5-15). But I couldn't find the problems.

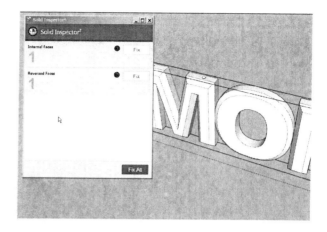

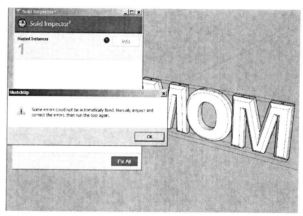

Figure 5-14 Solid Inspector[2] found flaws but couldn't fix them all.

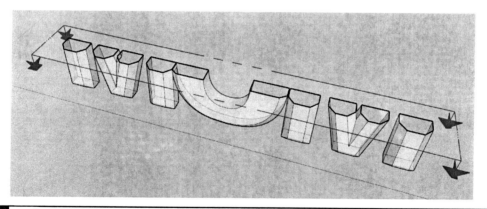

Figure 5-15 Make the model transparent, and use the Section tool to find flaws.

CleanUp³

This might be a job for CleanUp³! This tool does a lot of small things to optimize a model, such as purging unused items and erasing duplicate faces. Activate it at Extensions > CleanUp³, choose Clean to let it inspect everything, or choose a specific task (Figure 5-16). A pop-up box of options will appear. Check them all if you want, and then run the tool. It will inspect the model and return a box listing what it found (Figure 5-17). Click OK to let it fix them. In this case, CleanUp³ fixed the problems it found and returned another box saying that there were no more problems. However, the Entity Info box still shows the pencil holder as just a group, not a solid group. So let's try another tool.

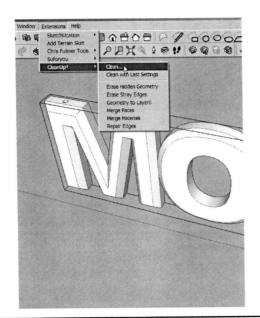 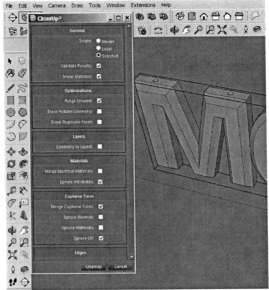

Figure 5-16 Using CleanUp³ to optimize a model.

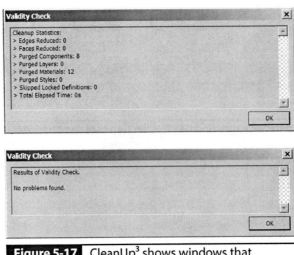

Figure 5-17 CleanUp³ shows windows that describe what was fixed.

SU Solid

SU Solid gives the model a more thorough inspection and offers additional information such as weight (useful for 3D printing because service bureaus charge by volume). Its toolbar (Figure 5-18) loads on installation. Run the mouse over each icon for a tooltip describing what it does. The free version accesses the Analysis and Delete

Single Edges tools; the other icons are grayed out because they are pay-version features. Highlight the model and then click on each icon to run its tool. Click on any SketchUp tool (one not on the SU Solid toolbar) to exit.

The Analyze tool returned the results in Figure 5-19. Zooming in, we see that the letters *M* and *O* overlap.

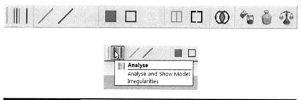

Figure 5-18 The SU Solid toolbar. Run the mouse over each icon to see what it does.

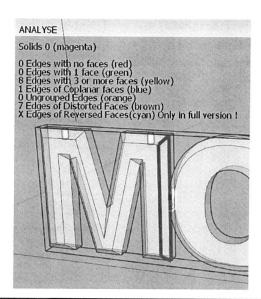
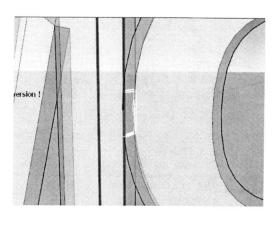

Figure 5-19 SU Solid's Analyze tool shows an overlap between two letters.

Manual Repair Techniques

Sometimes SU Solid fixes problems after you click its tools on the model multiple times. But, as with the other tools, you often have to fix the problems yourself or fix them enough for another go-round of the tool to work. So it looks like we have to figure out how to fix this overlap manually. The following troubleshooting techniques can be applied on any problematic model.

Move the Letters Apart

I tried moving the letters apart. SketchUp's "stickiness" made this impossible (Figure 5-20). In retrospect, I should have made a solid group of each letter while modeling it, which would have solved my current problem.

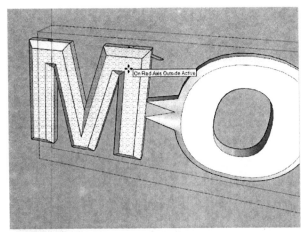

Figure 5-20 The letters stick together; hence, they can't be moved apart.

Intersect Faces

Maybe the Intersect Faces function will enable me to separate the letters? I selected the *M* and tried intersecting it with model (Figure 5-21) and then intersecting it with context. The results were marginal, so I selected the *O* and tried it again. No joy.

Split the Model into Parts

Perhaps splitting the model into parts will work (Figure 5-22). Insert the Section tool, right-click, and choose Create Group from Slice. All geometry at that exact location will get grouped (incidentally, you can slide that group out with the Move tool if you want a section at that location). Highlight the group, right-click, and choose Explode. This splits the model into two pieces, one in front of the slice and one behind, enabling you to erase one side or the other. I experimented with different locations, including one at the *M* and *O* intersection, but didn't get results I was happy with. This might be a job for Autodesk's 123D Meshmixer.

> Splitting the model in two with Section/Explode is also useful when the model is too big to fit on the printer's build plate. Break it in two in a location that makes the most sense (usually down the middle), and print the two parts as separate models. Then glue or pin them together.

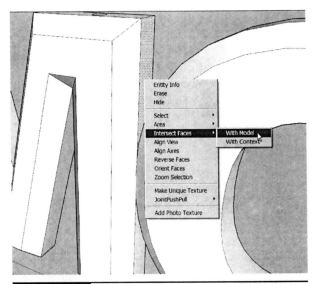

Figure 5-21 An attempt to fix the problem with the Intersect Faces function.

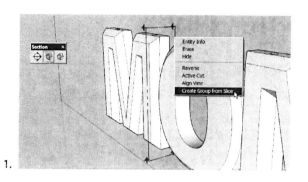

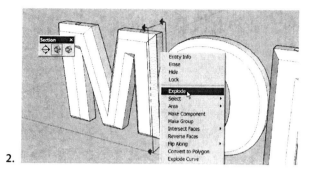

Figure 5-22 A model can be split into parts with the Section tool. The highlighted portion of the M is one part; the rest is another part.

Export the SketchUp Model as an .stl File

Before we can import this model into 123D Meshmixer, we need to export it as an .stl file

(Figure 5-23). Make sure that it's the size you want and that its units are in millimeters (mm). The Tape Measure can change both size and units. Click it on opposite ends of the model, type the size wanted, and add *mm* to the number

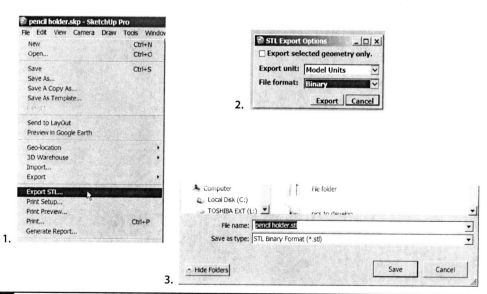

Figure 5-23 Export the model as an .stl file.

if the model's units are currently different. It's best to convert it to millimeters before printing because slicing software and printers use that unit, and it's the most accurate. Using millimeters will also enable you to scale the model precisely inside the slicer if you end up needing to scale it there.

The .stl extension we downloaded in Chapter 4 exports the model as an .stl file. Click on File > Export STL. A box will appear with Units and File Format options. Choose the model's units and Binary because it makes smaller files than the ASCII option. Then name and export the file. The model is now an .stl file, which is a mesh model inside a shell.

123D Meshmixer

Launch 123D Meshmixer. You'll see the screen in Figure 5-24. Click on File > Import, navigate to the .stl, and bring it in. Meshmixer immediately flags the overlapping letters with a red line (Figure 5-25).

To navigate, hold the SPACEBAR down. A "hotbox" will appear. Pan, Orbit, and Zoom are among its icons. To move or orient the model, press and release the T key on the keyboard. A manipulator will appear with arrows, angles, and planes all connected at an origin. Drag an arrow to move the model in the direction the

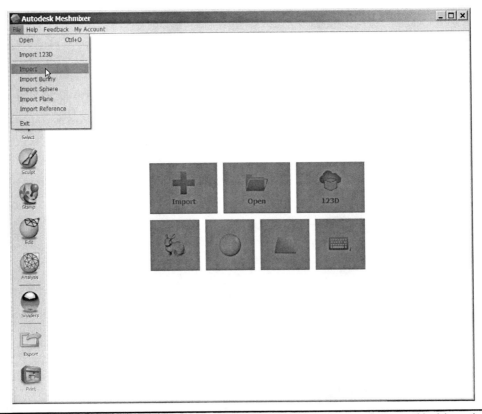

Figure 5-24 The 123D Meshmixer launch screen. At File > Import, navigate to the pencil holder .stl and import it.

Figure 5-25 The imported .stl file. 123D Meshmixer flagged the overlap with a red line.

arrow points. Drag an angle to rotate the model along that angle. Click and drag a plane to scale the model along that plane. Click and drag the origin to scale the model evenly in all directions.

Hold the cursor directly over the angle marks while rotating to snap in the marks' increments (Figure 5-26). Hit the UP/DOWN ARROW keys on the keyboard to increase/decrease the number of angle degree marks.

Click the Analysis icon (Figure 5-27) for a submenu of tools that check a model for 3D printability. Click Inspector at the top of the submenu. It flags all defects. Here apparently the only flaw is the overlap. We can fix it manually or use the Auto Repair All button. The latter is faster and easier but can have unpredictable results. I clicked it, and Meshmixer fixed the overlap by removing a portion of the letter. Because this can be easily fixed in SketchUp by deleting the now-partial *M* and copying/pasting the other *M* in its place, I'll simply export this Meshmixer file as an .stl file and import it into SketchUp. Click on the Export icon at the bottom of the menu bar. Note all the options. Choose .stl/Binary and we're good to go.

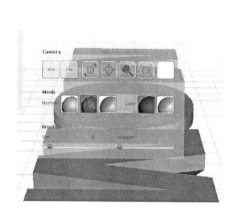

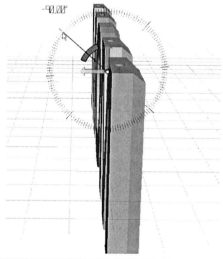

Figure 5-26 The hotbox on the left is accessed with the SPACEBAR. The manipulator on the right is accessed by pressing and releasing the T key.

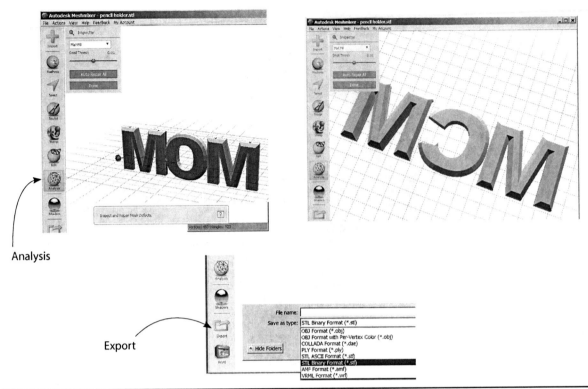

Analysis

Export

Figure 5-27 The Inspector tool flagged the defect. Clicking Auto Repair All resulted in the fix shown on the right. Export this 123D Meshmixer file as an .stl file to import into SketchUp.

Other 123D Meshmixer Analysis Features

The Analysis icon also contains the following tools, which you may want to run on your .stl files:

- **Units/Scale.** Click to set the model's units (they should be millimeters for printing). Scale the model by typing a number in one text field. The numbers in the other fields will scale proportionately. When you change units, a dialog box appears asking if you want to keep the numbers the same (e.g., 10" becomes 10 mm) or convert the number to the equivalent in the new unit.

- **Measure/Dimension.** Click here to measure features and distances between points. Options are type and direction; a numeric distance appears after clicking points.

- **Stability.** Check the surface area and volume of the model. A ball shows the model's center of mass; a green ball means that it is stable (will stand up), and a red ball means that it's unstable.

- **Strength.** Click this tool to see if the model is strong enough to be printed. Solid green means it is; red areas indicate weakness.

- **Overhangs.** Red areas show overhangs that need support during printing. Meshmixer can generate supports.

Tinkercad

A way to fix small .stl files is with Tinkercad. Point your Chrome browser to www.tinkercad .com (Figure 5-28). Log in with the same

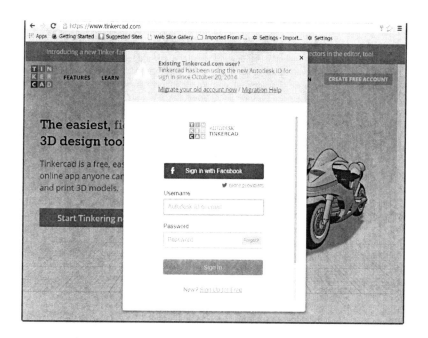

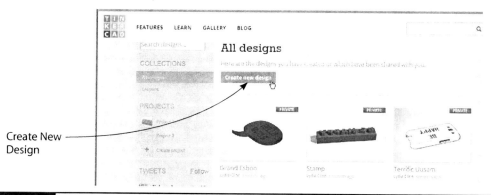

Create New
Design

Figure 5-28 The Tinkercad homepage and login screen (*top*) and personal homepage (*bottom*).

Autodesk account you used for Meshmixer, and you'll be taken to your personal homepage. This is where your Tinkercad designs are accessed. Click the Create New Design button (Figure 5-29). That takes you to the Tinkercad interface. Once there, click the Import bar. Next, click File to navigate to the .stl file, set the scale to 100 percent, choose the units the file was originally made in, and click the Import button (not the ENTER key).

On import, Tinkercad inspects and repairs any flaws. Know that it simply repairs them to make them printable. The repairs may affect the appearance in ways you don't want. However, you can download the repaired Tinkercad file as an .stl file and import it into SketchUp to see if the repairs make the file easier to work with. Figure 5-30 shows the imported pencil holder file. Tinkercad didn't change its appearance. Click on the color graphic in the upper-left corner. This takes you to your homepage, where there's a thumbnail of your project now. Tinkercad automatically saves files and changes made to them as you work. It names them with random words that you can rename.

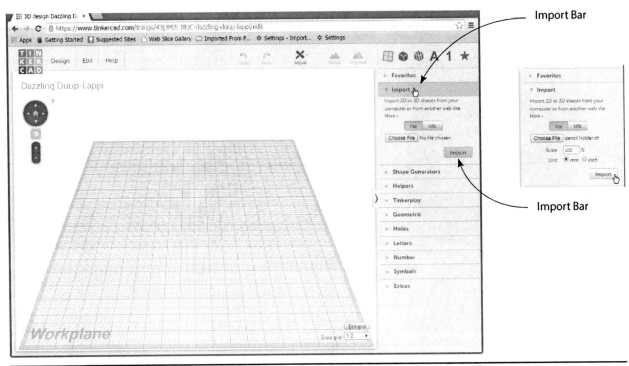

Figure 5-29 Tinkercad's interface. Click the Import bar and button to bring in the .stl file.

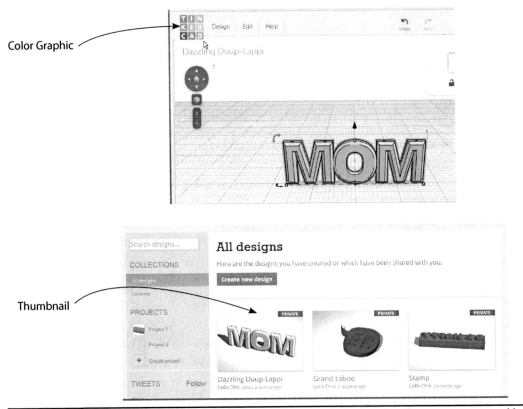

Figure 5-30 Tinkercad saves files as you work and places their thumbnails on your personal homepage.

To download this Tinkercad file as an .stl file, click on the thumbnail to enlarge it, and then click on the Download for 3D Printing button (Figure 5-31). Choose the .stl option and one will download to your computer.

Import the .stl file into SketchUp. It will import at the origin, and you'll probably need to click the Zoom Extents button to see it. I ran the Solid Inspector[2] tool on it. This time it isolated the problematic *M* and deleted it when I clicked Fix All (Figure 5-32). I was able to copy and paste the other *M* in its place.

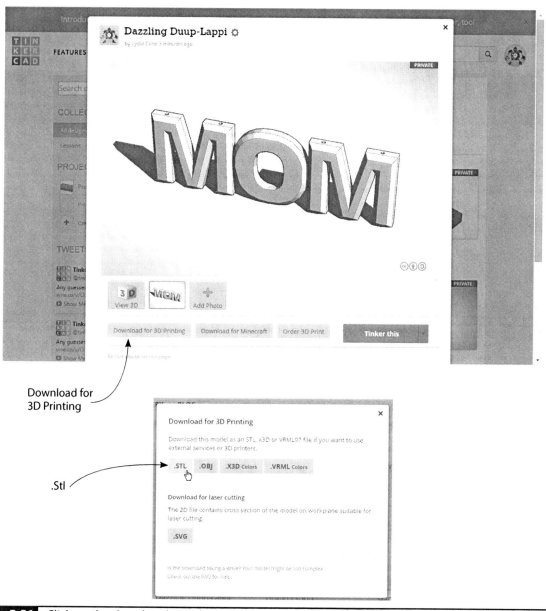

Download for 3D Printing

.Stl

Figure 5-31   Click on the thumbnail to enlarge it, then on Download for 3D Printing, then on the .stl option.

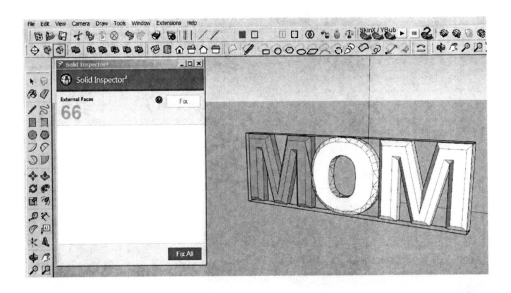

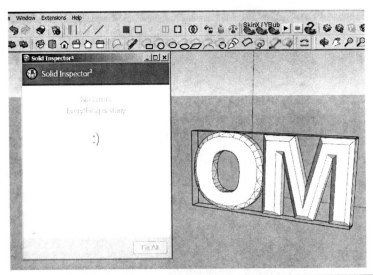

Figure 5-32   Solid Inspector² deleted the problematic *M*, and the other *M* was copy/pasted in its place.

Fixing the Dog Tag in 123D Meshmixer

Let's revisit the dog tag file, the version not fixed with Solid Inspector². It consists of multiple parts and hence is not 3D-printable. Depending on my workflow while modeling it, I could have ended up with a multi-part file, none of which were solid groups, and making them solid in SketchUp would be difficult. Let's see how to make the model a solid, 3D-printable file in Meshmixer.

Export the model from SketchUp as an .stl file, and then import it into Meshmixer. 123D Meshmixer flagged all the individual pieces as problems by outlining them in red (Figure 5-33). Click on the Edit icon and then on Solid in the submenu. A drop-down arrow offers Blocky, Fast, and Accurate choices; clicking Accurate gives the results shown.

Remember when we clicked on Analysis > Inspector to fix the pencil holder? The Analysis submenu contains more tools that optimize a file

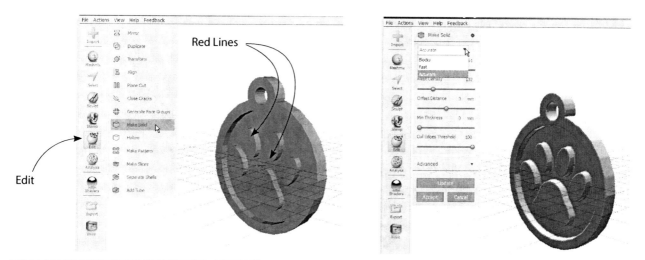

Figure 5-33 Click on Edit > Make Solid to weld all the dog tag's pieces into one solid piece.

for 3D printing (Figure 5-34). The Thickness tool checks for areas that are too thin to be printed (e.g., they might break when you remove the supports or during shipping). I ran the Thickness tool on the dog tag, and in Figure 5-35 you can see that it identified many areas of weakness. I moved the slider to left, which adaptively thickened those areas. This resulted in the dog tag turning completely green, meaning that it is now appropriately thickened. Click "Done."

Figure 5-34 The Analysis icon contains a menu of tools that optimize a file for 3D printability.

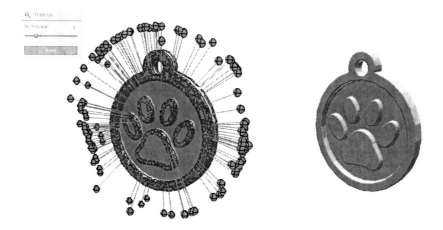

Figure 5-35 Inspector > Thickness identifies and fixes areas of weakness.

Analysis > Units/Dimensions lets you size the model, and Analysis > Strength colors weak areas yellow and red (Figure 5-36). Those often need to be adjusted back in SketchUp.

The Edit menu contains the Hollow tool, which removes all material inside a solid model with one click (Figure 5-37). The exact wall thickness is adjustable. A hollow model prints faster and uses less material, a particular concern when printing with service bureaus because they charge by material used. Know that even though a SketchUp model is empty like an air-filled balloon inside the SketchUp software, once it's exported as an .stl file and imported into Meshmixer, it becomes solid (that is, has a continuous volume).

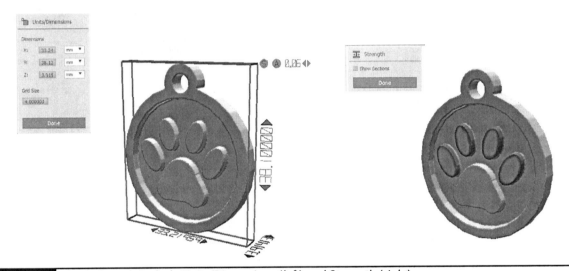

Figure 5-36 Analysis tools include Units/Dimensions (*left*) and Strength (*right*).

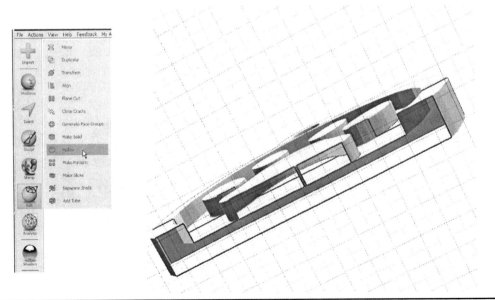

Figure 5-37 Click Edit > Hollow to make the model empty inside.

Coloring a Model

To date, we've only worked with uncolored models. Let's discuss how to add color and texture (pixel images) to a model and export it in a format that a 3D printer can read.

Applying Plain Colors

Click on the Paint Bucket (Figure 5-38). This opens a window of folders. Click on the Colors folder, click on a color and then click on the grouped fingernail we made in Chapter 4. You can color outside or inside a group's editing box. Clicking outside it colors the whole model, and you can change the color by clicking another color onto it. Only uncolored faces will be affected; if a face has already been colored inside the editing box, you'll need to open the editing box to change the color.

Coloring inside the editing box gives you more control because only the face you click on gets colored. To return to the default color, click the white/gray box in the panel's upper-right corner, and click on the model (Figure 5-39a). On a Mac, the default color is the first one in the Colors In Model list (Figure 5-39b). If you color a model, color all its front faces; don't leave the default color on some. Make sure that all the back faces have the default color.

Figure 5-38 The Paint Bucket tool.

(a)

Default Color

(b)

Scroll down to see colors

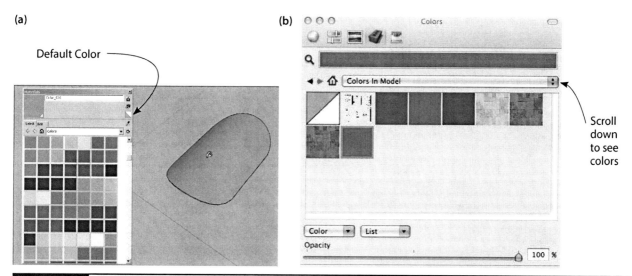

Figure 5-39 **Figure 5-39** (a) Click on a color, and then click on the model. The default color is the white/gray box. (b) The default color on a Mac is the first one in the Colors in Model list.

Applying a Raster File to a Flat Surface

You can dress your model up with an imported pixel file! Jpg, .png, .tiff, and .bmp files can be imported as a texture or image. The difference between texture and image in SketchUp is that a texture *tiles*, or repeats, and an image doesn't.

To import a raster file as a texture:

1. Click on File > Import. Navigate to the file, click the Use as Texture button, and click Open. Place the file on an existing plane with two clicks (Figure 5-40). You can't place it on a group, so before importing it,

either explode the group or open the file's editing box.

2. Adjust the tile size. Click on Window > Materials. Then click on the House icon, which shows all textures in the model. Click on the imported texture and then on the Edit tab. A panel will appear; if you don't see all the fields shown in Figure 5-41, drag the bottom of the panel down with the mouse to make it larger. The numbers in the text fields are the tile length and width. Highlight one and type a different size. The other will adjust proportionately. To change the length and width non-proportionately, click the link icon next to the tile fields first.

1. Click File > Import.

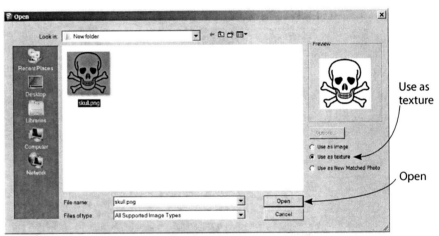

2. Navigate to, and select the file. Click Use as Texture and Open.

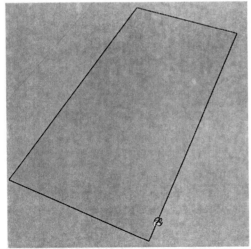

3. Click twice onto a piece of loose geometry to color it with the pixel file.

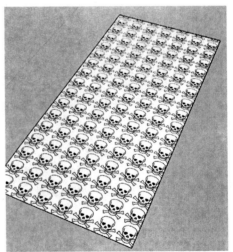

4. The texture repeats according to its tile size.

Figure 5-40   Importing a pixel file into SketchUp.

Edit Tab

House Icon

Tile Adjustment Fields

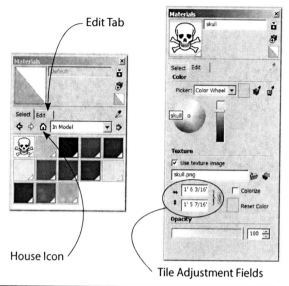

Figure 5-41 Click on the House icon and then on the Edit tab to find the tile adjustment fields.

New Tile Number

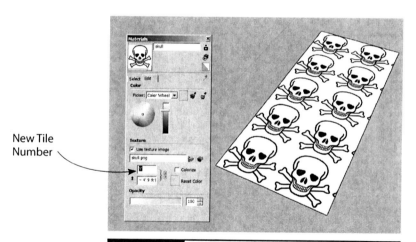

Figure 5-42 Type a new tile size to change the number of repeats.

Experimentation is needed to find an appropriate tile size. Figure 5-42 shows the image with a larger tile size.

To import a pixel file as an image, click on File > Import (Figure 5-43). Navigate to the file, click the Use as Image button, and press ENTER. Place it on a plane with two clicks. Note that it doesn't tile; there's just one picture. You can import it directly onto a plane or anywhere in the work space. Here I've drawn a rectangle around it and pushed/pulled it into a box.

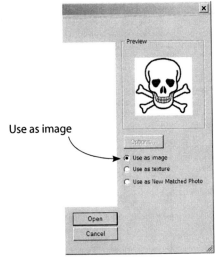

Use as image

1. Import the image.

2. Click twice to place.

3. Outline with a rectangle.

4. Push/pull.

Figure 5-43 Importing a file as an image results in one picture of it.

Applying a Raster Image to a Curved Surface

Textures can be applied to curved surfaces but images cannot. So if the file is an image, turn it into a texture. Select it (click on its perimeter to do this), right-click, and choose Explode (Figure 5-44). We'll apply this texture to the fingernail with the following steps:

1. Highlight the texture, right-click, and choose Texture > Projected (Figure 5-45).

2. "Sample" the texture (Figure 5-46). Activate the Paint Bucket. On the color panel, click the Select tab. On a PC, click the eyedropper, click it onto the texture, and

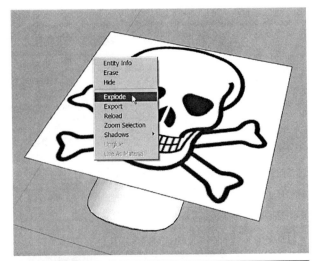

Figure 5-44 Explode an image to turn it into a texture.

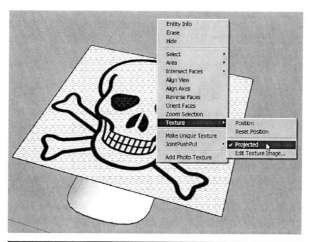

Figure 5-45 Right-click a highlighted texture, and choose Texture > Projected.

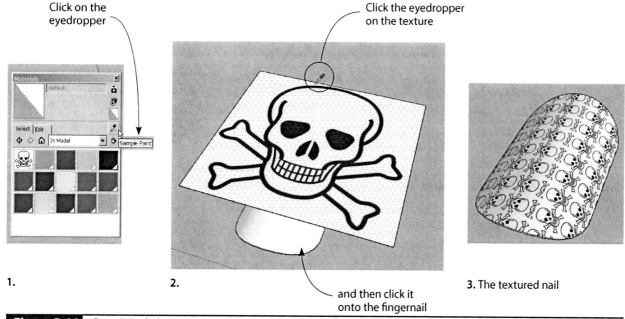

Click on the eyedropper

1.

Click the eyedropper on the texture

2.

and then click it onto the fingernail

3. The textured nail

Figure 5-46 On a PC, click on the eyedropper to sample the file. On a Mac, use the Paint Bucket while holding the COMMAND key down.

then click it onto the fingernail. The Mac doesn't have an eyedropper; instead, hold the COMMAND key down, click the Paint Bucket onto the file, release the COMMAND key, and click the Paint Bucket onto the fingernail.

Different imported file sizes tend to affect the tiling size. In Figure 5-47 I managed to sample

and scale one repeat of the design onto the nail. If a design doesn't position the way you want, try rotating the file before you import it. You can also right-click on the texture, choose Texture > Position, and drag the image around with the mouse (Figure 5-48). Move (don't drag) the colored pins to reposition, rotate, warp, and scale the design.

Figure 5-47   The design with just one visible repeat.

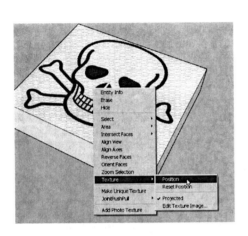

Figure 5-48   At Texture > Position, move the pins to alter the design's appearance.

Face Orientation

The front side of a polygon must face out for a model to be solid. The front side is also the side that gets painted. Orientation is especially important for color models because if the front face is reversed, the printer can't read it. Know that when multiple faces are selected, the side on which the Paint Bucket clicks is the side that gets painted on all faces. So if you paint the front of one face, all the selected faces' fronts get painted.

If you paint the back of one face, all the selected faces' backs get painted. Furthermore, if you select and paint a front face plus its edges, all the model's edges get painted. If you select and paint a model's back face and edges, no edges get painted.

If designs and colors don't show up, click on View > Face Style, and check Shaded with Textures. If Monochrome is checked, only the white and gray default colors will appear.

To see painted edges, click on Window > Styles, click the Edit tab, and then click the Edge Settings button. Click the drop-down menu in the Color field, and choose By Material.

Purge the Model

Large imported files slow SketchUp down, so it's good practice to reduce their size in digital imaging software before importing them. Unused components and materials also slow the model down because they remain in it even after deletion. At Window > Model Info > Statistics, click the Purge Unused button to get rid of all unused items (Figure 5-49). Purge the model again right before exporting it into a 3D-printable file format.

The 3D-Printer Facsimile Template

When you're done modeling, scale the model to the size you want it in millimeters (click the Tape Measure on opposite ends, and type the units followed by *mm*). Millimeters is the most accurate unit, and it's the one that most 3D printers recognize.

To ensure that your model will fit in your 3D printer, go to Window > Preferences > Template and choose 3D Printing–Millimeters (Figure 5-50). Close the file, reopen it, and that template is now current; a facsimile of a 3D printer appears. Alternatively, access this facsimile at Window > Components, and look in the Components Sampler folder.

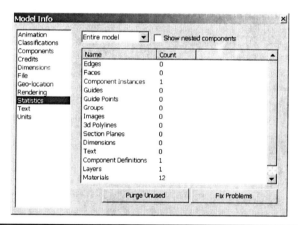

Figure 5-49 Purge unused components and materials to keep the model's size down.

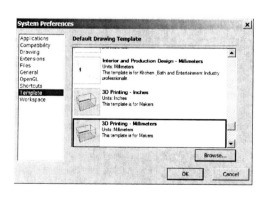
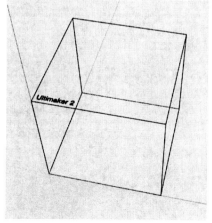

Figure 5-50 At Window > Preferences > Template, choose 3D Printing–Millimeters. Then close and reopen the file. A facsimile of a 3D printer will appear; here an Ultimaker 2 is shown.

The facsimile is actually a dynamic component, which is a component that is coded to do things. Many components in the Warehouse are dynamic; they're indicated by this symbol: . Click on View > Toolbars > Dynamic Components to activate the Dynamic Components toolbar (Figure 5-51).

The Dynamic Components toolbar has three icons (Figure 5-52). The first is the Interact tool; if you hover it over a selected dynamic component and it glows, this means that clicking on the dynamic component will activate something. For example, a cabinet door will open or a drawer will roll out. The second icon has the component options. I scrolled to MakerBot Replicator 2, clicked the Apply button, and the dynamic component changed to that printer. The top graphic in Figure 5-53 shows a MakerBot Replicator 2 facsimile. The bottom graphic shows the Chapter 4 dog tag model inside MakerBot Desktop slicing

software, from where it's printed. Both the facsimile and the slicing software let you see if the model fits inside the printer.

If you don't see your printer in the options listed, choose one whose size is similar. SketchUp Pro users can directly adjust the facsimile size via the third icon on the Dynamic Components toolbar. Once you've ensured that the model fits inside the printer, delete the dynamic component.

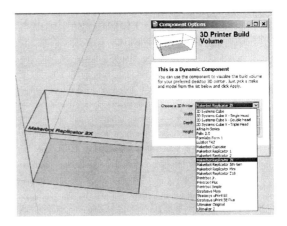

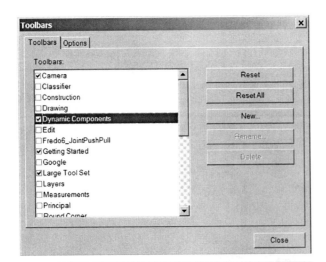

Figure 5-51 Activate the Dynamic Components toolbar.

Figure 5-52 The Dynamic Components toolbar.

Figure 5-53 Select the component, scroll to your printer, click Apply, and the dynamic component will change. The facsimile should be the same size as the facsimile inside the slicer software that ultimately sends the model to the printer.

Export the Colored Model

Stl files don't contain color information. You need to export the model as an .obj file (Figure 5-54), a .wrl file (a Pro feature), or a .dae file (a Make feature; look for the COLLADA option). In the process, an accompanying .mtl file and possibly a .jpg file, will be generated containing the color information. All those files are needed to print in color or to import the model and its color into another software program. Export them to a folder made specifically to hold them and keep them all together there.

If you're wondering what the difference between .obj, .wrl, and .dae files are, an .obj file is like an .stl file but with color. A .wrl file exports color, transparency, light, and reflections. Both are readily importable into many software programs, which you might want to use to develop the model further. A .dae file contains color information, too, but it's more of an exchange file, which is one that can be read by various programs. The commercial machines owned by service bureaus can read all of them, so if all you want to do is 3D print the model, all three formats are fine. However, if you want to import the file into another program, more will accept an .obj or .wrl format than a .dae format. For instance, Meshmixer imports .obj files but not .dae files.

Let's head to Chapter 6 now, where we'll learn how to print these models up.

Summary

A model must meet specific requirements to be 3D-printable. It's not enough for it to simply be in an .stl format. Third-party tools and programs such as Solid Inspector[2] and 123D Meshmixer are available to inspect and help fix them. The more tools a modeler knows, the more likely it is that he or she will be able to quickly and successfully print a model.

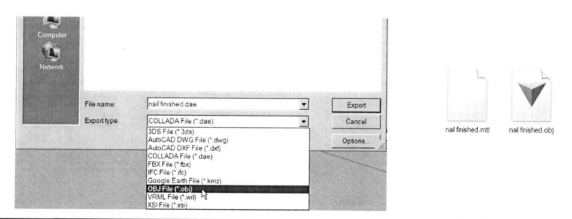

Figure 5-54 Export a color file as an .obj file. Keep it and the material files generated with it in the same folder.

Resources

- *3D Printing with Autodesk 123D, Tinkercad, and MakerBot*, by Lydia Sloan Cline (TAB/McGraw-Hill, New York, 2015).

- **Lydia's Meshmixer playlist.** Has videos that show how to use the app: https://www.youtube.com/playlist?list=PLij ALmDbL90-rJP3bSTgTHyk3_guqcHk3.

- **3D Printing Blog:** http://3dprinting-blog .com/.

- **Design tips from Shapeways:** http://www.shapeways.com/materials/strong -and-flexible-plastic.

- **Free software that converts many file formats:** http://meshlab.sourceforge.net/.

- **Official 123D Meshmixer forum:** http://meshmixer.com/forum/.

CHAPTER 6

3D-Printing the Model

IN THIS CHAPTER WE'LL DISCUSS how to print
models that have been checked for printability
and are ready to go. We'll cover the 3D printing
process, features of consumer-level printers,
slicer software, and troubleshooting. We'll also
look at printing with a service bureau.

Types of 3D Printing

3D printing encompasses the different ways a
physical model is built from a digital one. The .stl
or other printable format file is a mesh inside an
invisible shell. That mesh consists of hundreds
or thousands of polygons and their X, Y, Z
coordinates. The printer uses that information
to create the physical model layer by layer. This
is called *additive manufacturing*, meaning that
material is added to create the item.

The three most common 3D printing
processes are

- **Fused deposition modeling (FDM).** Successive
  thin layers of a material are melted and
  placed on a platform to physically create
  the digital model (Figure 6-1). This is how
  consumer-level printers operate.

- **Stereo lithography (STL).** This process gave
  the .stl file format its name. It creates a model
  layer by layer with a photosensitive liquid
  plastic that's hardened with an ultraviolet
  light beam. The resulting print is of higher
  quality than an FDM print.

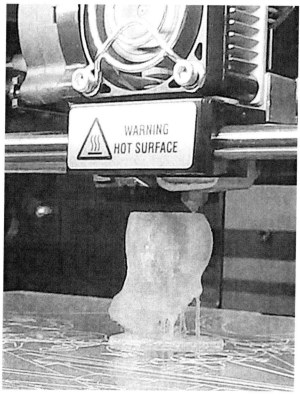

Figure 6-1 In the FDM process, thin layers of a
material are built up to create the form
of the digital model.

- **Selective laser sintering (SLS).** This is similar
  to the STL process but uses powder melted
  with a laser beam. It enables printing with
  metal, which is not possible with the FDM
  and STL processes. It's the most expensive
  process, and machines that do it are owned
  by companies that make commercial-level
  prints.

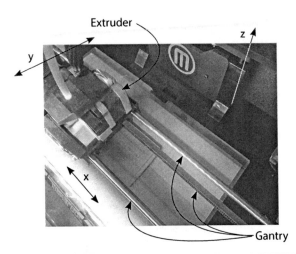

Figure 6-2   The extruder and gantry. The extruder moves along the *x*, *y*, and *z* axes to deposit filament.

Figure 6-3   Filament is a long string of material that is wound around a spool. String diameter and spool size varies.

Printers have an *extruder*, or printer head, which melts the material and deposits it onto a *build plate* or printer bed. The extruder moves in three directions: left to right along the *x* axis, forward and backward along the *y* axis, and up or down along the *z* axis. It's attached to a *gantry*, which is a system of rods and belts that moves along a frame (Figure 6-2).

Filament

The FDM process uses *filament*, which is a long string of material that is wound around a spool (Figure 6-3). Its standard diameter is 1.75 mm (the most common) or 3 mm. Filament comes in solid, translucent, fluorescent, glow-in-the-dark, and sparkly colors. Post- (after) production, most types can be primed, painted, sanded, machined, drilled, and glued. This enables you to print large models in separate parts and then glue or pin them together (Figure 6-4).

Consumer printers use two types of thermoplastic: PLA and ABS. PLA (polylactic acid) is corn-based and biodegradable. It's appropriate for beginners because it's easy to work with. PLA is best for small, simple household items. It is used on an unheated build

Figure 6-4   These 3D-printed parts were drilled and fastened together.

plate and cools quickly, which helps to avoid warping. It has a low melting point, so you have to keep the printer on a relatively low heat or the model will deform. The final product is shiny. ABS (acrylonitrile butadiene styrene) is a petroleum-based material that emits strong fumes. It's strong, flexible, and appropriate for detailed pieces with joints and interlocking pieces. It requires a heated build plate, and its high melting point means that the model won't deform under high heat. Because it takes longer to cool, however, the model may warp in the process. It is not a beginner-friendly material.

Other filament types include

- TPE (thermoplastic elastomer), which is as flexible as rubber. NinjaFlex is a common brand.

- Laywood, which looks like wood.

- Laybrick, which looks like stone.

- PVA (polyvinyl alcohol), which supports intricate models and dissolves in water.

- HIPS (high-impact polystyrene), which supports intricate models and dissolves in limonene, a citrus-scented chemical.

- Polycarbonate, which is very strong.

- Nylon, which absorbs color after the model is printed.

- Translucent. Taulman T-glase is a common brand.

- Conductive ABS, which conducts electricity.

ABS and PLA filaments are food safe, but the additives that give them properties such as fluorescence and translucence are not. Models made with the FDM process have porous surfaces that collect bacteria, so they should be coated before use. Some filaments, such as T-glase, cannot withstand dishwasher heat. All have their challenges to work with, such as how well they stick to a build plate, what temperature they must be heated to, how quickly they cool, how much they shrink, how well they bridge (span openings), and which printers they can be used in.

Store filament in airtight containers with desiccant packets. Moisture causes it to expand, which will jam the extruder (some Makers bake moisture-exposed filament at a low heat in an oven for a couple of hours to dry). Dirt on filament also will jam the extruder. Filament quality varies widely among manufacturers and should be bought only from reputable ones. Low-quality filament has nubs and uneven widths that can cause extruder damage. Use a digital caliper (Figure 6-5) to measure filament to ensure that it's the same thickness throughout. Because bad parts tend to be lengthy, you can

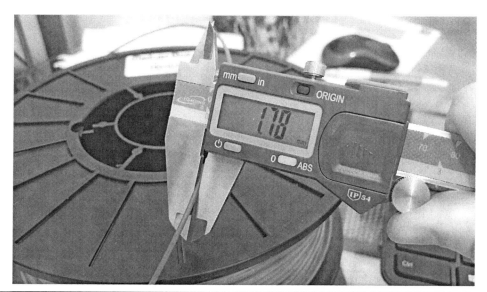

Figure 6-5   Use a digital caliper to check filament widths.

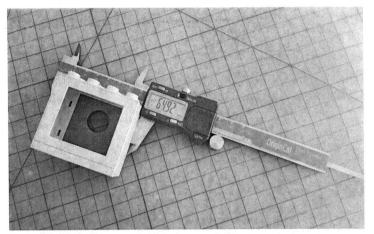

Exterior Width

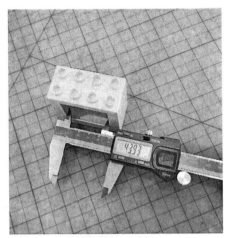

Interior Width

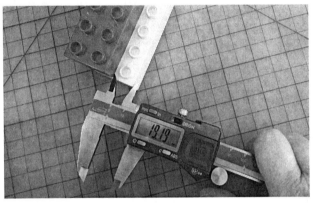

Step Width

Height

Figure 6-6 A digital caliper measures exterior and interior widths, steps, and heights.

measure random locations on the filament. It's also useful to measure items for which you're making accessories (Figure 6-6).

Choosing a Home 3D Printer

Shopping for a consumer printer? You've got choices! One is to buy the parts and assemble it yourself. The many varieties of the open-source RepRap is the most famous example of this. Another is to buy a kit and assemble its parts. These options offer the greatest cost savings and ability to customize. If you want a printer that is already assembled, popular manufacturers include LulzBot, Afinia, Cubify, FlashForge,

MakerBot, Ultimaker, and Formlabs. That said, all printers require some work, such as manually positioning the build plate.

Even though FDM printers work the same way, their physical appearance and features are different (Figure 6-7). For example, some have enclosures around the print area; others don't. Some have WiFi capability and onboard cameras. Some have dual extruders (Figure 6-8), which enables using two filaments (e.g., two colors or one color and one dissolvable). Build plates can be made of glass or plastic, removable or nonremovable, heated or unheated, and different sizes. Some printers have slots for SD cards; some don't. Some use standard-size

Figure 6-7 Price, features, and appearance of 3D printers vary. (*Left*) An Oni (www.onitechnology.com); (*right*) a MakerBot (www.makerbot.com).

Figure 6-8 Dual extruders enable printing with two different filaments.

or let you service it yourself. Some do not allow self-service as a warranty condition.

You need to have an idea of what you'll be printing to select appropriate features and filaments. Look at the output of different printers to judge quality and level of detail (Figure 6-9). There are also online communities built around specific brands and types. Google to find them. You'll be impressed at how Makers are always pushing their machines' limits, trying new things, and sharing the results.

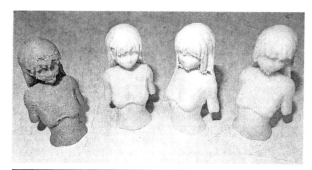

Figure 6-9 These models were made with the same settings on four different printers.

filament spools; others use a proprietary size, meaning that you can only purchase them from the manufacturer. Some printers accept multiple filament materials, and others accept only one. Some are customizable, let you upgrade parts,

Printing and Slicing Software

3D printers operate from printing and slicing software. Printing software—also called front-end software—contains options for the model's physical settings and a control panel while the printer is running. Slicing software—also called back-end software—divides the model into the many layers that the printer builds one by one and develops the G-code that the printer reads. Some software incorporates front- and back-end programs, making the file preparation process easier. Popular free programs include Cura, Skeinforge, Slic3r, ReplicatorG, and Repetier-Host. Simplify3D is a good for-pay one. Microsoft's Windows 8.1 platform has 3D-printing support. Brand-name printers come with software written for them, but anyone can download it for free and likely be able to use it on a different machine. For example, Cura is written for the Ultimaker, but it can be used on other machines. All slicers offer different options, settings, and adjustments, some more than others. Makers all have their personal favorites.

Firmware

Firmware runs the printer itself and is pre-installed on the motherboard. It gets updates just like the software. Check the website of the company that made the printer for updates. If you're using the printer's own software, you may get push notices of firmware updates. There also may be a menu entry that searches for updates (Figure 6-11).

Print Detail and Quality

Print detail and quality are factors of the machine and its slicer settings. Here are settings that affect the print:

- **Resolution.** This is the thickness, or height, of each layer it creates, measured in microns (Figure 6-12). Thin layers provide more detail than thick ones. The thinner the layer, the lower is the resolution number; 100 microns is a paper-thin layer that produces nice detail.

  While some machines do better jobs and can accommodate more settings, there will always be visible layer lines. Post-processing work such as sanding and tumbling is needed for a higher level of smoothness. This might be done by machine but is usually done by hand.

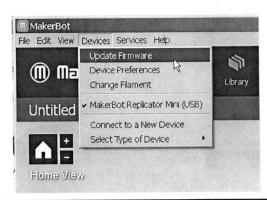

Figure 6-11   The MakerBot's firmware is updated through MakerBot Desktop software.

Figure 6-12   The model on the left has a lower resolution than the model on the right.

▪ **Nozzle opening diameter.** The nozzle is the spout at the bottom of the extruder head (Figure 6-13). Nozzles are available in different diameters. A small nozzle creates thinner (and slower) layers and hence more detail. A wide nozzle creates thicker (and faster) layers and hence less detail. Note the text on the luggage tag in Figure 6-14. It was printed with a 0.4-mm nozzle, which was still too wide to do a good job on the *R*. Choose simple fonts because flourishes and serifs print poorly.

> SketchUp uses the fonts in your computer's operating system. If you want a different font, download and install it in your operating system and it will appear in SketchUp's font choices.

▪ **Infill.** This is the amount, or density, of material. You can adjust the slicer settings to fill between 0 and 100 percent. A 0 percent infill creates a hollow model. The left graphic in Figure 6-15 shows a slice through the luggage tag. Note the hollow between the border walls. The right graphic shows the border while being printed.

Figure 6-13 Extruder nozzles are available with different diameter openings.

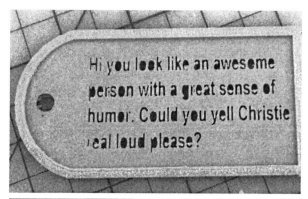

Figure 6-14 A luggage tag. Only simple fonts should be printed, and best results are obtained with a small nozzle.

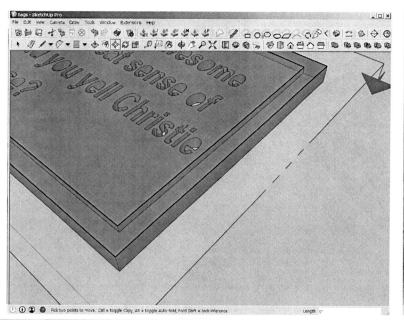

Figure 6-15 The SketchUp model (*left*) and the printed model (*right*). Note the hollow border.

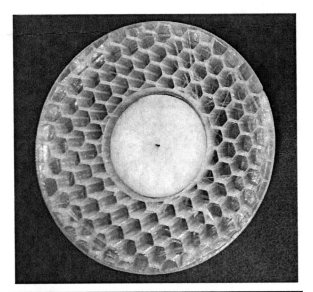

Figure 6-16   The honeycomb pattern of this candle holder is a 10 percent infill. (*Courtesy Susana Chen, High School of Community Leadership, New York, NY*)

Cross-hatch Infill

Figure 6-17   The cross-hatch pattern in the Chapter 4 dog tag is a 40 percent infill.

When the infill is greater than 0 but less than 100 percent, the printer makes a pattern, such as a lattice or honeycomb (Figures 6-16 and 6-17), to bridge the space. Every model requires a different infill (Figure 6-18). For example, lacy models with

thin walls need 100 percent infill because a lower percentage would result in weak walls. Chunky models work well with an infill between 10 and 40 percent. A 100 percent infill on a chunky model is a poor choice because it will create heat-dissipation

Figure 6-18   Lacy models (*left*) need a higher infill than chunky ones (*right*). (*Phone model courtesy of Susana Chen and Nazrul Akeem, High School of Community Leadership, New York, NY*)

problems during the printing process, possibly resulting in warping.

- **Print speed.** This is how fast the extruder moves. The higher the speed, the faster it will print. Fast printing produces less detailed prints than slow printing. A 1" × 1" model with a 0.4-mm nozzle and 10 percent infill takes about 25 minutes to print on a consumer machine. Large, intricate models can take hours or days to print.

- **Shells.** A shell is the wall around the model's perimeter. Most slicers make two shells by default. If an item crushes while under construction, increase its shell number. However, too many shells (typically over five, but depending on the model's size) may change the model's appearance.

- **Supports.** These are thin vertical layers of filament created to hold up *overhangs*, which are parts of the model with empty space below (Figure 6-19). Supports are needed for anything angled less than 45° to the horizontal; 0. 3-mm-diameter supports spaced every 1 cm are typically enough. Post-production, remove supports by snapping or cutting them off with pliers or a craft knife or by dissolving them (if you used dissolvable filament to make them).

Add supports after the model is scaled. If you rescale the model once it's in the slicer, the supports may not scale correctly with it and end up too small to be effective or too large for easy removal. If there are a lot of overhangs, the print may come out fluffy or droopy; consider breaking the digital model into parts for printing separately and then gluing or pinning the parts together. You can break a SketchUp model with the Section tool, as discussed in Chapter 5.

- **Raft.** This is a latticed, peel-away layer at the bottom of the model and at the bottom of supports to help them adhere to the build plate (Figure 6-20). Its layers print thicker than the model's layers. Putting a raft under a long, flat model also prevents its edges from curling if it cools unevenly during printing.

- **Bridge.** This is a horizontal support stretched between gaps to support construction above the gap.

Figure 6-19 Supports hold up parts of the model with empty space below.

Figure 6-20 A raft is a thick, peel-away lattice that helps a model adhere to the build plate. The blue surface is painter's tape.

Position on the Build Plate

Orient the model to require the least number of supports (Figure 6-21). This saves time and filament and avoids the scars that snapped-off supports leave. The slicer suggests an orientation based on how you've placed the model inside the printer facsimile, so give that placement some thought. Try to put the longest surface flat on the build plate. Also try to analyze where stress will be placed on the printed model (such as at a connection), and orient the model so that the filament grain runs parallel to that location. Otherwise, stress points will be held together simply by adhesion between the layers.

Adherence to the Build Plate

Models need to stick to the build plate to print successfully, and you'll generally need a craft spatula to remove them. If the model doesn't adhere to the build plate, here are some tips:

For both heated and unheated build plates:

- Make sure that the build plate is level, meaning that it is parallel to and the correct distance from the extruder.

- Use a raft.

- Cover the build plate with painter's tape or Kapton (polymide) tape (Figure 6-22). Use large pieces that cover the whole plate, not

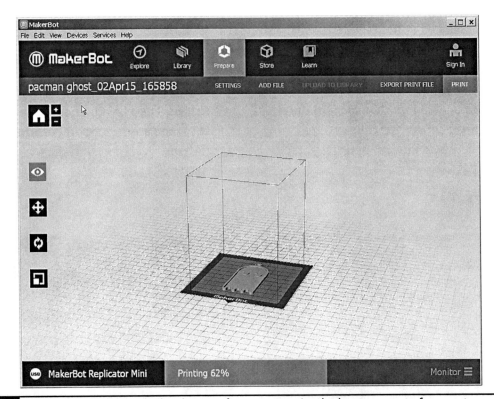

Figure 6-21   Orient the model inside the slicing software to require the least amount of support.

Figure 6-22 Kapton tape covers this build plate.

narrow strips. There must be no bubbling or wrinkling. Painter's tape has a texture that helps adhesion, and Kapton is smooth, which results in a smooth surface. Make the tape last longer by printing models in different locations on the build plate.

- Clean the bed, especially if you've used tape on it. A smooth surface with no dust, dirt, oil, or adhesive residue is needed. Clean all dust and debris out of the printer, too. Canned air works well for this.

- Slow down the extruder speed.

- Try a different filament, as you may have a bad batch. Research whether the filament is supposed to adhere to the surface you have it on. For example, flexible filament adheres to acrylic but not to painter's tape or glass.

- Replace an acrylic plate with a glass plate, as acrylic may warp after heavy usage, impeding adherence.

For a heated build plate:

- Wipe the build plate with a mixture of acetone and dissolved ABS filament. Let it dry before printing.

- Apply a very light coating of hairspray over the taped or untaped bed. Let it dry before printing. Clean with a damp rag afterward, and reapply another light coat before the next print.

- Clean the bed, especially if you've used hairspray on it, which leaves a gummy residue.

- Ensure that you're using the manufacturer's recommended temperature.

Troubleshooting

Problems are an expected part of 3D printing. Typical ones include

- **The nozzle and/or extruder clogs.** Know how to disassemble the extruder (Figure 6-23) and have the tools to do so (Figures 6-24 and 6-25). Soaking the nozzle in acetone dissolves ABS. Figure 6-26 shows tools to help remove filament clogs. Consider the nozzle a consumable and replace it often for best printing results. They can be bought cheaply in packs on Amazon.

- **Filament leaking outside the nozzle.** This results in burned glops deposited onto the model (Figure 6-27). Wrapping plumber's tape around the nozzle shaft as shown in Figure 6-28 helps prevent leaks. Also, extruder parts get vibrated loose after dozens of printing hours, and need to be tightened as general maintenance.

Figure 6-23 A disassembled extruder.

Figure 6-24 Useful tools for working on a printer.

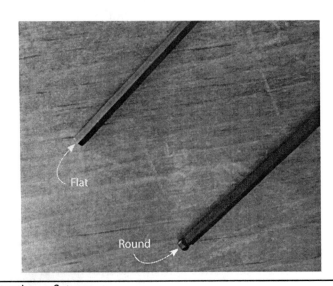

Figure 6-25 A round-head wrench works better than a flat one.

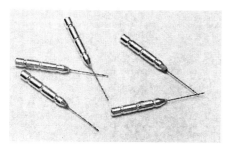

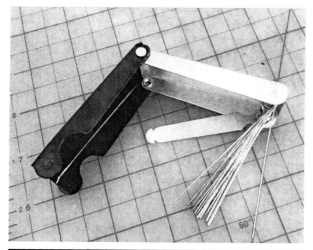

Figure 6-26 Drill bits help clean out clogs. (*Top*) For nozzles. (*Bottom*) Longer ones for reaching into the extruder.

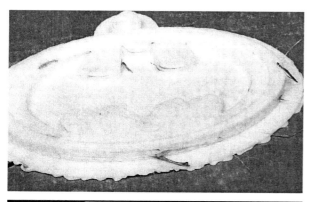

Figure 6-27 Filament leaking outside the nozzle caused these burned glops.

Figure 6-28 Wrap plumber's tape around the nozzle shaft to help prevent filament leaks.

- **The slicer doesn't connect to the printer.** Sometimes you need to set everything up in a certain order. Try this:
  1. Plug in the USB cord.
  2. Turn the printer on.
  3. Wait for a verification message that your computer has installed the drivers.
  4. Wait for the printer's chime or light to verify that it's ready to start.
  5. Open the slicer software.
  6. Open an .stl file.

- **The slicer takes an overly long time to send the model to the printer.** Is the model resting directly on the facsimile platform? Slicers can still process a model not on the platform (Figure 6-29), but it often takes much longer.

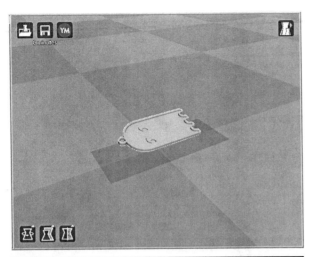

Figure 6-29 The model is not resting directly on the facsimile platform. This may result in a longer processing time.

- **There are gaps between the printed layers** (Figure 6-30). Try any of these:

    1. Adjust the temperature, speed, or layer height settings. However, be aware that you may damage your printer if you deviate significantly from recommended manufacturer settings.

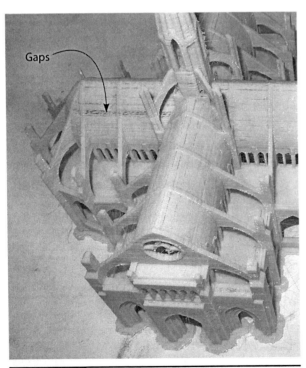

Gaps

Figure 6-30 Gaps between the printed layers.

2. Install a new nozzle.

3. Re-level or clean the build plate. Place a level on the plate to see whether it's perfectly horizontal (Figure 6-31).

4. Readjust the tension on the filament and/or the belt.

5. Lubricate the printer rods. This should be done every 50 hours of use anyway. Use a PTFE (polytetrafluoroethylene)–based grease.

6. Try a different slicer software.

Bead will center between the two vertical lines when the level is horizontal.

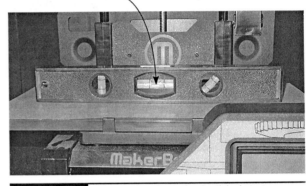

Figure 6-31 Place a level on the build plate to see whether it's perfectly horizontal.

Odds and Ends

Learn your slicer's options, such as settings (Figure 6-32) and positioning (Figure 6-33). Know the specific procedure for loading and unloading filament and how to pause the machine, which is useful for changing an empty spool mid-print or changing colors.

Adjust the computer's energy settings so that it doesn't fall asleep while printing. Some slicers will halt the printing when this happens, and you may not be able to restart it.

When unsure whether something can be printed, print a small, quick sample. Alternatively, send small, cheap models to a service bureau to study the results.

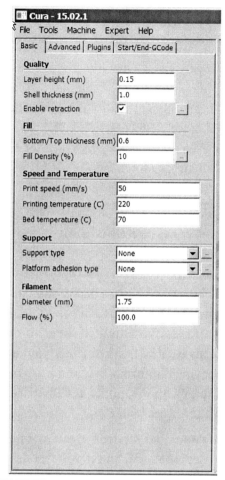

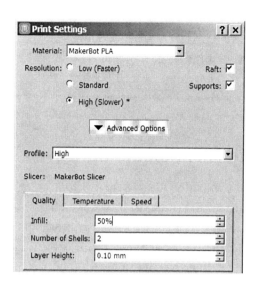

Figure 6-32 Settings options on Cura (*left*) and MakerBot Desktop (*right*).

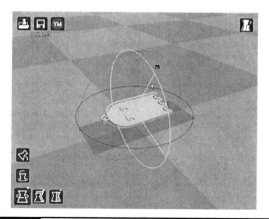

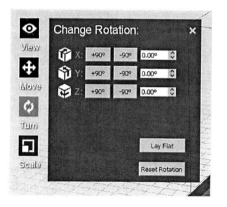

Figure 6-33 Repositioning the model in Cura (*left*) and MakerBot Desktop (*right*).

Place the printer on a sturdy table that supports it without shaking. It should be in a well-ventilated area that is free from drafts. Printers are kind of noisy, so keep this in mind when choosing a location, too.

Don't leave a printer unattended for long lengths of time while a model is under construction. This can pose a safety hazard if something overheats or otherwise goes wrong.

Printing from an SD Card

A slicer can export the file in a format for copying onto an SD (memory) card, which you can then insert directly into the printer (Figure 6-34). This is advantageous for several reasons. One is that the communication between the computer and printer can fail at any time, which won't happen if it's reading directly from an SD card. So this is particularly useful for long printing times. Another advantage is that printing from the computer may cause other programs to operate slowly. Finally, printing directly from a card enables you to turn off or disconnect your computer while printing.

Printing with a Service Bureau

Service bureaus are companies that offer 3D printing services. They use commercial-grade machines and have more filament choices, hence you'll get a higher quality product. They also accept more file formats than just .stl and .obj; some even take raw .skp files. See the formats one bureau, Sculpteo, accepts at http://sculpteo .com/en/help/#accepted-formats.

Upload your model to the service bureau's website, either directly or through the slicing software if it has that option. You don't need to include supports. After you choose a material, you'll be quoted a price based on material and volume. If the model is not printable, you'll get an error message, although some bureaus may make an attempt to fix it. Enter your address and payment information, and if no problems are found with the model, it will arrive in the mail a few weeks later.

Be aware that service bureaus are almost entirely automated, and if your model isn't perfectly clear about what you want, assumptions may be made by the service bureau's own software and printers that may not

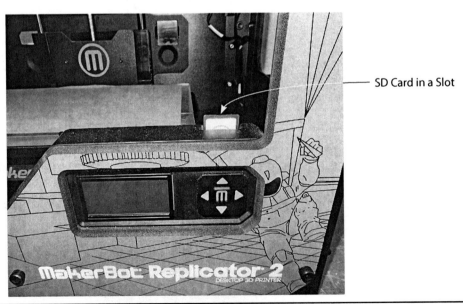

SD Card in a Slot

Figure 6-34 An SD card and slot for printing directly through the machine.

be what you intended. For instance, a model of a building whose designer's intent was to simply show form and nothing inside may get filled solid if you don't offset the perimeter walls to make them a specific thickness. Because you're charged by amount of material used, this adds needless cost.

Some service bureaus offer post-processing services, such as removing the supports (it's not a given that a third party will automatically remove them), polishing, or coloring.

The most popular service bureaus are shapeways.com, sculpteo.com/en, ponoko.com, and i.materialise.com. There's also 3dhubs.com, a network of local printers who may help to tweak designs as well as print them. Check out local Maker spaces, which often have helpful people, or even your public library, which may have a 3D printer or two.

Summary

The physical printing process can be as challenging as the digital modeling one. In this chapter we discussed printer features, types of filament, the physical 3D printing process, slicer software, troubleshooting, and how to print through a service bureau.

Resources

- **Shapeways printing tutorials:** http://shapeways.com/tutorials?li=nav

- **Website showing prints gone bad:** http://epic3dprintingfail.tumblr.com/

- **3D printer comparisons:** http://www.inside3dp.com/reviews/3d-printer-comparison/

- **3D Printing magazine:** http://www.3ders.org/

- **3D Printing Blog:** http://3dprinting-blog.com/

- **Compare printers:** http://3dforged.com/best-3d-printers/

- **Facebook groups:** 3D Printing Experts, 3D Printing Club, 3D Printing

- **How to measure with a digital caliper:** https://www.youtube.com/watch?v=EqF6I0NHPm0

- **How to level a build plate:** https://www.youtube.com/watch?v=LzgcFGbMxMU

CHAPTER 7

CNC Fabrication with Pro, 123D Make, and Cut2D

IN THIS CHAPTER we'll turn SketchUp models into scaled 2D files that can be fabricated on a computer numerical control (CNC) router. This manufacturing method is common for making crafts, terrain models, and furniture. It's a more mature technology than 3D printing, and the post-processing work is often easier than cleaning a print. CNC machines operate with computer-aided manufacturing (CAM) software.

Specifically, we'll slice up the terrain model from Chapter 4 with a CAM program called Autodesk 123D Make and export it into a CAM-appropriate format. We'll use SketchUp Pro to export the Chapter 3 nameplate into scaled .pdf and .dxf formats and then import the .pdf file into a CAM program called *Vetric Cut2D*. In the process, we'll learn about CAM software, the files it needs, how to generate them, and a bit about the machines themselves.

What's a Router?

A *router* is a machine that cuts or carves sheets of wood, plastic, glass, nonferrous metal, foam, and wax (Figure 7-1). Routers that are larger and can cut harder metals are *mills*. Both routers

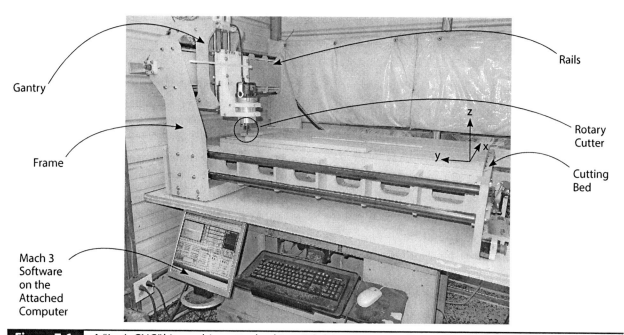

Gantry

Frame

Mach 3 Software on the Attached Computer

Rails

Rotary Cutter

Cutting Bed

Figure 7-1 A "Joe's CNC" kit machine attached to a PC. (*Hammerspace, Kansas City, MO*)

Figure 7-2 End and ball mills come in thousands of shapes and sizes.

Figure 7-3 CNC machines cut the ornaments (*left*) and carved the Mayan calendar (*right*).

and mills consist of a metal frame that holds a gantry, rotary cutter (Figure 7-2), and cutting bed. A "sacrificial sheet" is placed under the cutting sheet to protect the bed.

Routers use end mills and ball mills (Figure 7-2). These cut side to side and move along the x, y, and z axes as opposed to drills, which just cut up and down. Hence, they can carve as well as cut (Figure 7-3). A CNC router holds and cuts the material via inputted computer code as opposed to an operator manually holding and manipulating the material against a cutter.

> Nonferrous metals are the "softer" metals such as brass and aluminum. Foam, wax, and medium-density fiberboard (MDF) are materials used to make patterns and test designs before committing to more expensive materials.

CNC fabrication is a subtractive process, meaning that material is removed to make an object. This process generates a lot of chips. Another CNC cutting tool is a *laser*, a hot beam of light that cuts a material cleanly (Figure 7-4). Depending on the project, you may want to use a router or a laser.

Figure 7-4 Lasers cut a material cleanly. The lower-right corner was lasered and sand blasted to remove the burn color.

Computer-Aided Manufacturing Software

Computer-aided manufacturing (CAM) software creates the files that run a CNC machine. Expensive commercial machines like those made by Haas have built-in computers that use proprietary software. Kit and lower-cost machines attach to a PC through a USB port and run either software specifically written for them or generic cross-platform software.

CAM programs vary greatly in features, cost, and complexity. Some have rudimentary modeling, drawing, and editing capabilities; create cutting patterns; and generate *G-code*, which is the language machines understand (Figure 7-5). Some have one or two of those functions, and some just read G-code. Some CAM programs read 3D .stl files, but most require scaled 2D vector files in a .pdf, .dxf, or .eps format. Popular CAM programs include Fusion 360 and 123D Make (Autodesk), V-Carve Pro and Cut 2D (Vetric), Mastercam (CNC Software), Mach 3 (Artsoft), and Cambam. Good, free CAM software is not plentiful like 3D printing software.

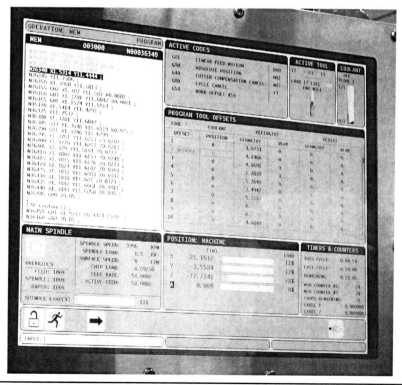

Figure 7-5 G-code displayed on the built-in computer of a Haas CNC mill.

The Process for CNCing with SketchUp

Here's an overview of the process for CNCing with SketchUp:

1. Model the project. Run Solid Inspector², CleanUp³, SU Solid, and/or any other tool you learned about in Chapter 5 to remove extra geometry and optimize the model as much as possible. A SketchUp model doesn't have to be a solid group or component to import into a CAM program, but if it has a lot of defects, such as nonmanifold edges or holes, it may import with errors or be difficult to work with (Figure 7-6).

2. Export the SketchUp model as an .stl, .pdf, or .dxf file depending on the requirements of the CNC program you plan to use it with. The .stl extension in the Extension Warehouse works with both Make and Pro. However, only Pro exports a file as a .pdf or .dxf file. Figure 7-7 shows Pro's export choices.

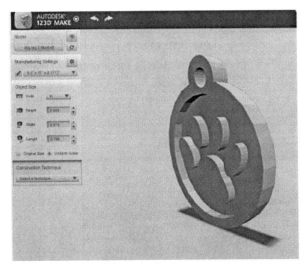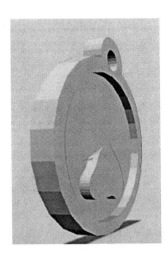

Figure 7-6   The dog tag (*left*) was fixed with Solid Inspector² and imported successfully into 123D Make. The unfixed tag (*right*) imported with parts missing.

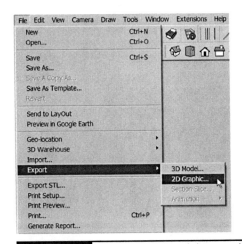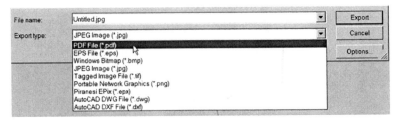

Figure 7-7   Exporting a model as a .pdf file in SketchUp Pro.

SketchUp Make users can obtain .pdf and .dxf files with a bit more work. Free and low-cost .pdf writers exist; two popular ones are at cutepdf.com and pdf995.com. Once one is installed, click on File > Print, and print to the .pdf writer instead of to the paper printer. It will generate a .pdf file with a default name that's the same as the model name. Be aware that all .pdf programs do not export vector files. Pdf files can also be raster (pixel) files, and these are not compatible with CAM software. A way to tell a vector from a raster file is by opening it with Adobe Reader and clicking on it. Raster images will highlight as when clicked on, and their lines will look jagged when enlarged greater than 400 percent.

A way to get a .dxf file from SketchUp Make is to export the model as a .dae file, import it into Blender (a free modeling program at blender.org), and export it from Blender as a .dxf file. A .dxf format is always a vector file.

3. Import the .stl, .pdf, or .dxf file into a CAM program that does what you need, such as generate pattern pieces on sheets and/or generate G-code files.

4. Import the G-code files into a CNC machine that reads and cuts them out, or upload them to a service bureau such as ponoko.com or shapeways.com.

Architectural Terrain Model Project

Let's turn the Chapter 5 terrain model into pattern pieces. First, export it from SketchUp as a .stl file (Figure 7-8). Then download Autodesk 123D Make at 123dapp.com/make,

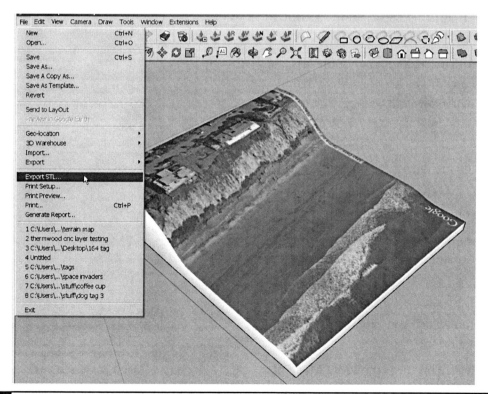

Figure 7-8   Export the SketchUp terrain model as an .stl file.

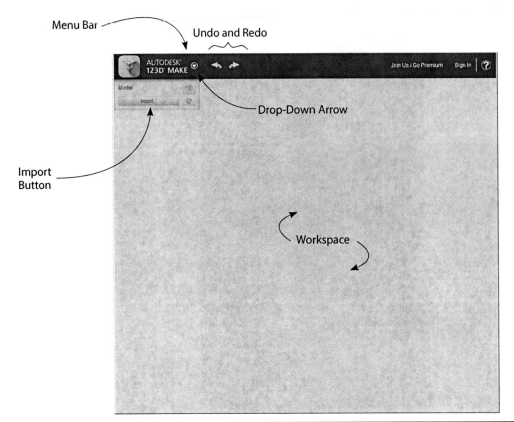

Undo and Redo

Drop-Down Arrow

Import
Button

Workspace

Figure 7-9 The 123D Make interface.

and launch it. Figure 7-9 shows the 123D Make interface.

There's a workspace, Menu bar at the top, and vertical panel on the left. The Menu bar has a drop-down arrow that accesses a submenu (Figure 7-10), and next to it are Undo and Redo arrows.

The submenu includes these functions:

- **New.** This closes the current file and makes the Import button appear.

- **Open Example Shapes.** This brings up choices of premade files with which to experiment.

- **Open.** This brings up a navigation browser window from which you can open an existing 123D Make project.

- **Export Mesh.** This turns the 123D Make file into an .stl or .obj file.

Figure 7-10 Click the drop-down arrow in the Menu bar to see this submenu.

Terrain .stl
Button

Manufacturing
Settings

View Cube

Navigation
Panel

Figure 7-11   Import the terrain model .stl file.

Click the Import button to bring up a navigation browser. Then navigate to and import the terrain model .stl file (Figure 7-11). Some new items now appear: a terrain .stl button (click it to close the file and open a navigation browser), manufacturing settings, a View Cube, and a navigation panel.

Navigating the 123D Make Interface

You can move around the interface three ways: with the View Cube, the mouse, and the Navigation bar.

▪ **View Cube.** This shows the model's orientation on the workspace. Click the mouse on it and drag to rotate. The model will rotate with it. Click on the View Cube's sides to see the model as a top, front, or side view. Hover the mouse over the View Cube to make a house icon appear. Clicking on that house returns the model to the default position and perspective view.

▪ **Mouse.** Right-click anywhere on the screen, and drag the cursor to tumble the model (move it at any angle around the workspace). Press the scroll wheel down to drag the cursor, which pans, or slides, the model around the workspace. Roll the scroll wheel up and down to zoom in and out.

▪ **Navigation panel.** Click on the first four icons to tumble, pan, zoom, and fit (fit fills the screen with the model). Click on the last icon to toggle between perspective and paraline views.

Choose the Manufacturing Settings and Object Size

We need to select the sheet size from which the project will be cut. In the Manufacturing Settings box, click the drop-down menu, and scroll through the choices (Figure 7-12).

Click on the pencil for a screen that shows all settings choices (Figure 7-13). It has graphics at the bottom for adding your own presets, which is handy for non-standard materials. Click the plus sign to add a new preset. To make a variation of a preset, click that preset to highlight it, and then click the double plus sign to make a duplicate. Enter the new settings. To delete a preset, highlight it, and click the minus sign.

Now select the physical size in the Object Size panel (Figure 7-14). Choose its units, and adjust the height, width, and length, if needed. Clicking the Uniform Scale button scales the file in all directions. When it's unclicked, you can scale the model differently along the three axes. Clicking the Original Size button reverts the file back to its size at import. The larger the file, the more

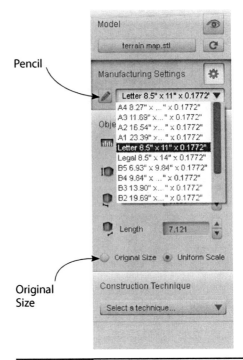

Pencil

Original Size

Figure 7-12 Click the pencil to access the Manufacturing Settings options.

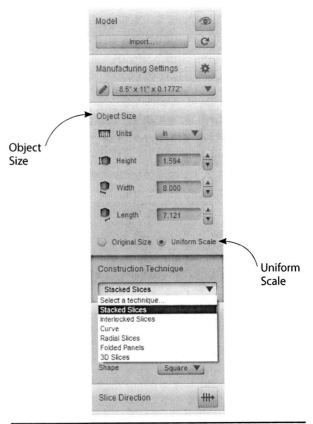

Object Size

Uniform Scale

Figure 7-14 Click the drop-down arrow to select a construction technique.

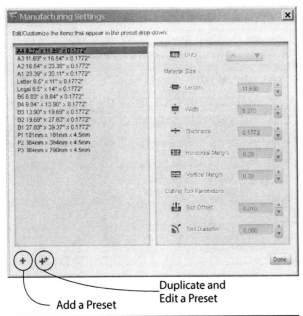

Duplicate and Edit a Preset

Add a Preset

Figure 7-13 Add your own presets in the settings choices window.

slices will be generated, and the more sheets you'll need.

Construction Techniques

At the bottom of the panel is the Construction Technique box. Click the drop-down arrow to select how to fabricate the model.

There are six techniques: Stacked Slices, Interlocked Slices, Curve, Radial Slices, Folded Panels, and 3D Slices. The model will update to reflect each technique you choose, with accompanying 2D graphics of slices laid out on sheets. All construction techniques have options with which you can finesse the model. Most overlap, but each has at least one option that is unique to it. As you finesse the options, the slices automatically update. Know that *slices* and *parts* are not synonymous. A slice can consist of multiple parts.

The 123D Make Process Using Stacked Slices

We'll go through the whole 123D Make process on the terrain model now, from choosing the technique to printing a .pdf file.

1. Click on the Stacked Slices technique. This makes cross-sectional slices for gluing and stacking on top of each other (Figure 7-15), which is how physical terrain models are made. Note the graphics under the Cut Layout tab.

   The Dowels option (Figure 7-16) creates pegs that help to align and hold the slices together. You can choose their size, location, and shape. Move a dowel by highlighting and dragging it; remove it by highlighting and hitting the DELETE key. Click on the Slice Direction icon to make a blue handle appear that you can drag to change the slice direction (Figure 7-17). Blue parts are

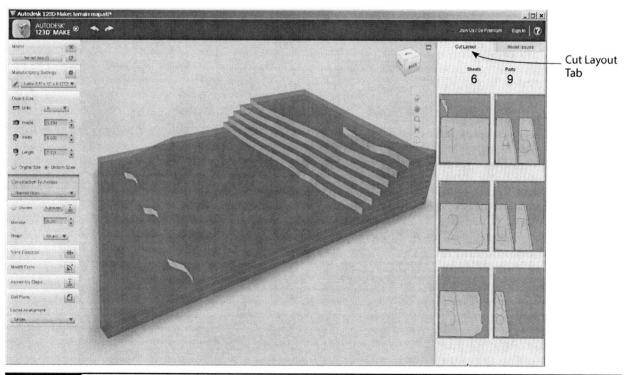

Cut Layout Tab

Figure 7-15 The Stacked Slices option turns the model into cross-sectional slices.

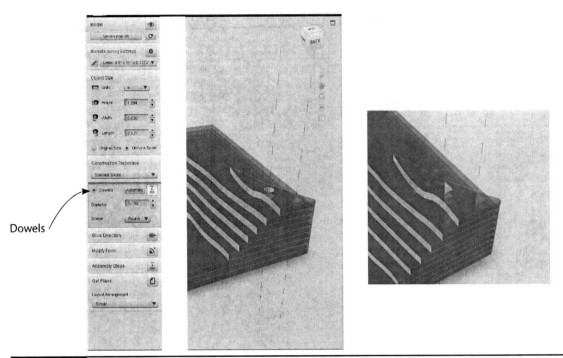

Dowels

Figure 7-16 Adjust dowel locations by dragging them. Adjust their size and shape via the menu.

Blue Handle

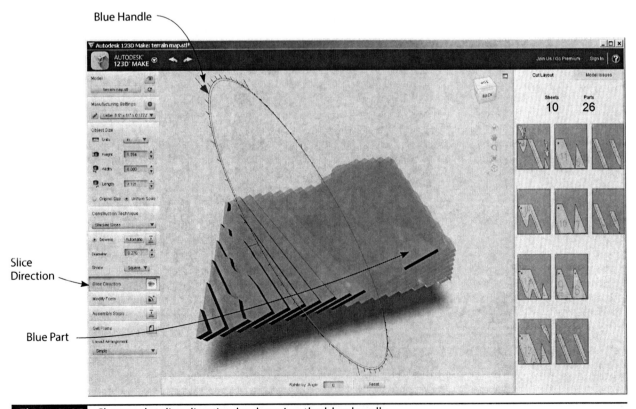

Slice
Direction

Blue Part

Figure 7-17 Change the slice direction by dragging the blue handle.

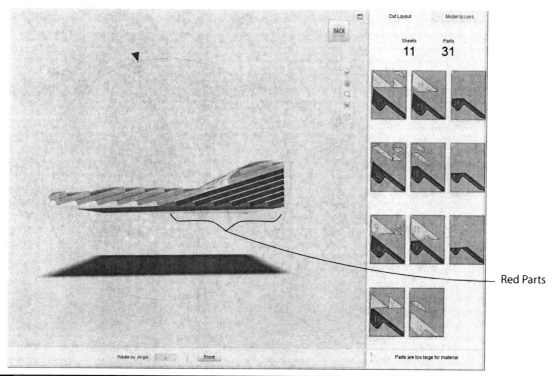

Figure 7-18 Red parts are too large for the cutting sheet.

unconnected to others. Red parts are too large for the cutting sheet (Figure 7-18). Click on the Model Issues tab (Figure 7-19) for a list of problems. Note that when selecting options, you need to click on the icon; nothing happens if you simply click the larger button the icon is on.

Even if everything looks fine, problems may appear when you export the model to a file. If this happens, click on the Modify Form icon to bring up three buttons at the bottom of the screen: Hollow, Thicken, and Shrinkwrap (Figure 7-20). Hollow shells out the model, reducing the amount of material needed to build it. Thicken widens cutouts that are too thin to print. Shrinkwrap approximates and smoothes details that are too fine to cut out and closes holes. Select one, adjust its slider, and click Done.

Be aware that Shrinkwrap affects the whole model, not just a problem slice. So the end product may change the model's

Figure 7-19 Click the Model Issues tab to see any problems. None are shown here.

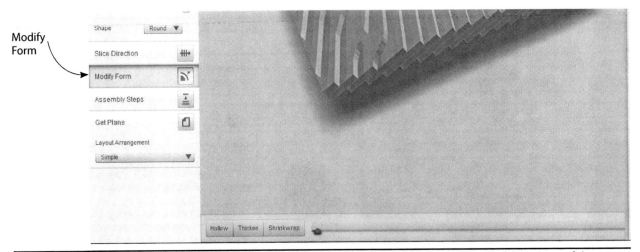

Figure 7-20 The Modify Form icon accesses Hollow, Thicken, and Shrinkwrap options at the bottom of the screen.

appearance. Additionally, know a machine cutter's limitations; for example, a router can't make square corners. Appropriate clearances, which are spaces for parts to connect, are also needed.

2. Click on the Assembly Steps icon and choose the material (Figure 7-21). Drag the mouse along the slider at the bottom of the screen to watch an animation that puts the parts together (Figure 7-22). To view an assembly sheet, click on it in the Assembly Reference panel (Figure 7-23). Move the sheet by clicking and dragging; scroll the mouse wheel to zoom in on a part; click the X in the upper-right corner to return to the model.

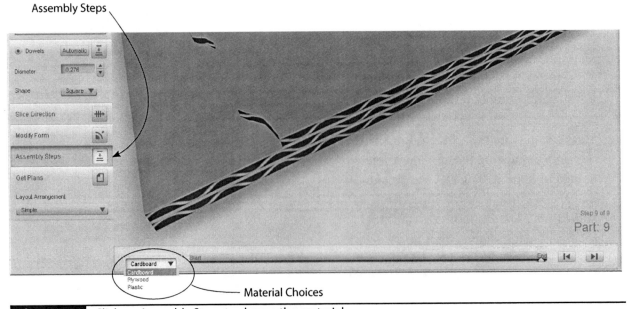

Figure 7-21 Click on Assembly Steps to choose the material.

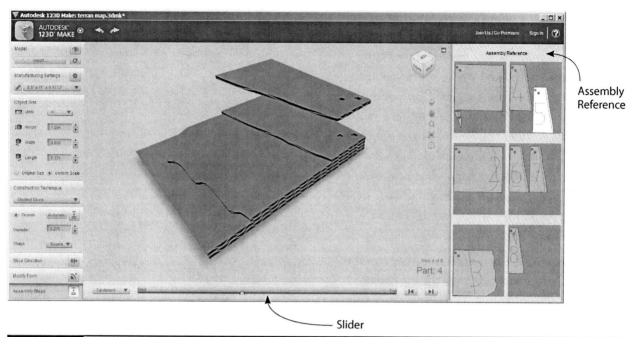

Figure 7-22 Move the slider to view an animation.

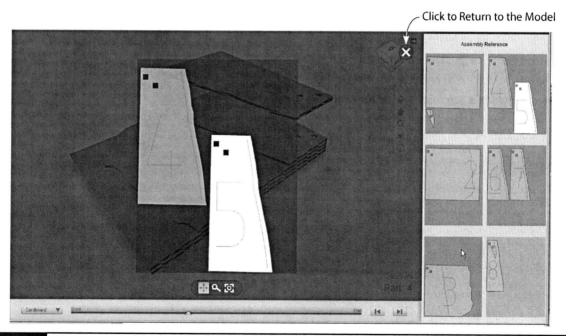

Figure 7-23 View an assembly sheet by clicking on its graphic.

Study the graphics in the Cut Layout tab (Figure 7-24) because they show how many sheets of material are needed, how many slices and parts, and where problems are. Hyphenated labels, such as Z-3-2, describe the part's axis (Z), the slice number (3), and the part number (2). Blue outlines are the model's outside edge, green outlines are cuts inside the model for hollows, yellow outlines are scored guides for placing parts during assembly, and red parts have too many errors to be built.

The parts are arranged to use as much of each sheet as possible, and all are oriented in the same direction. Currently, there is no other way to arrange them. If you want to change the arrangement—perhaps a wood grain faces the wrong direction—import the file into another vector program such as Inkscape (free at inkscape.org) or Adobe Illustrator, and change it there.

3. Export the file. Click on the Get Plans icon, choose the format (.eps, .pdf, or .dxf) and units, and click Export (Figure 7-25). The file will be exported into a format that is ready for postprocessing (G-code generation) in any CNC machine. The .eps option creates a zip folder with separate files for each sheet. It and .dxf files put text on one layer and profiles on another, which is

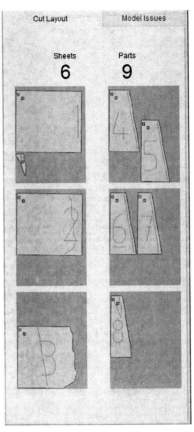

Figure 7-24 The graphics in the Cut Layout tab show the material sheets.

useful when importing into another CAM program for CNC cutting.

The .pdf option creates one file that puts all the slices on separate pages (Figure 7-26).

Figure 7-25 Click on the Get Plans icon, choose an .eps, .pdf., or .dxf file format, and export it.

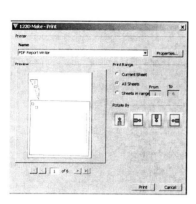

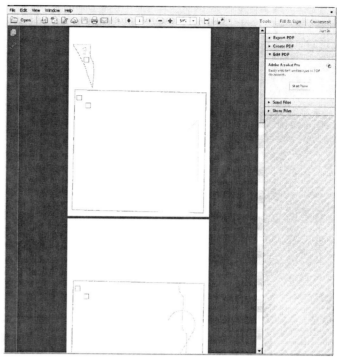

Figure 7-26 The .pdf export screen and the resulting file.

Other 123D Make Construction Techniques

You probably noticed that the Construction Technique field has multiple options. Click the head in the Open Example Shapes submenu (Figure 7-27), and let's apply the other techniques to it (Figure 7-28).

- **Interlocked Slices.** This slices the model into two stacks of slotted parts that lock together in a grid. It uses less material than the Stacked Slices technique.

- **Curve.** This cuts slices perpendicular to a curve.

Open Example Shapes

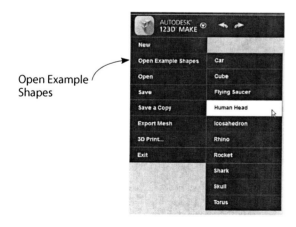

Figure 7-27 Click the head in the Open Example Shapes submenu.

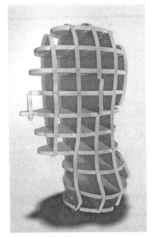
Interlocked Slices

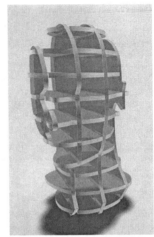
Curve

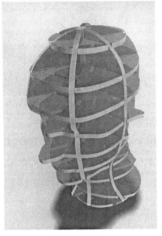
Radial Slices

Folded Panels

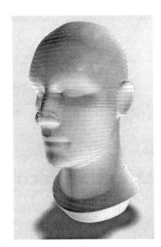
3D

Figure 7-28 Construction Technique options applied to the example head.

- **Radial Slices.** This creates radiating slices from a central point. It works best on round, symmetrical models.

- **Folded Panels.** This turns the model into 2D segments, or panels, of triangular meshes that are folded multiple times and held together with tabs (Figure 7-29).

- **3D.** This slices the model and follows its form as opposed to merely stepping them to create a form, as with the Stacked Slices technique.

Figure 7-30 shows the fingernail from Chapter 5 with the Radial Slices and Interlocked Slices techniques. Both show red and blue parts. You may be able to fix them by finessing the

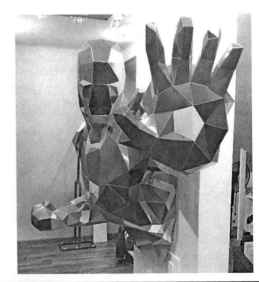

Figure 7-29 A folded panels art piece by dtworkshop.com.

Figure 7-30 The Radial Slices (*left*) and Interlocked Slices (*right*) techniques on the fingernail from Chapter 5.

technique's options. However, every technique won't work on every model.

Exporting a .pdf File from SketchUp

We can get scaled orthographic (2D) views directly from SketchUp, and here's how. First, change the default view from perspective to paraline by clicking on Camera > Parallel Projection. Then open the Views toolbar. Combine the Parallel Projection setting with a standard view (Figure 7-31) to generate a 2D drawing. Because standard views are aligned

along the red, green, and blue axes, the model must be aligned with those axes, too.

Change the workspace so that the background will print solid white (Figure 7-32). Go to Window > Styles to bring up the Default Styles dialog box. Under the Select tab, the Default collection folder should be visible in the text field; if it isn't, scroll to it. Click on the Construction Documentation style. The background will turn completely white. Then click on the View menu and uncheck Axes to remove the axes from the display (Figure 7-33).

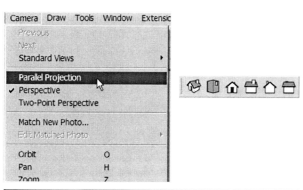

Figure 7-31 Combine Parallel Projection with the Views icons to generate orthographic views.

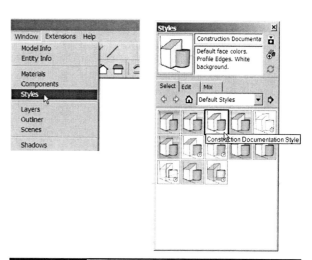

Figure 7-32 Change the style to Construction Documentation.

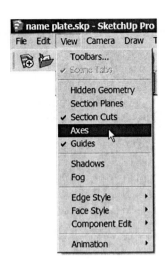

Figure 7-33   Turn off the axes display.

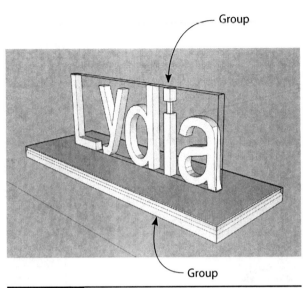

Figure 7-34   Group the name and base separately.

Name Plate Project

Let's export the name plate from Chapter 3 as a scaled, orthographic .pdf file using the .pdf writer that comes with SketchUp Pro. As mentioned earlier in this chapter, you can download and install a free .pdf writer to work with SketchUp Make. Just be aware that it may not have the same features as SketchUp Pro's .pdf writer and that you have to ensure that it exports vector files.

Group and Arrange the Parts

We're going to cut the letters and base separately. So group them separately to make them easier to lay out (Figure 7-34). Ideally, each will be a solid group; if not, run tools on them to make them so, or to clean them up as much as possible.

Lay the two groups flat and click on the Top View icon (Figure 7-35). We only need this view because depth information is entered manually into a CAM program.

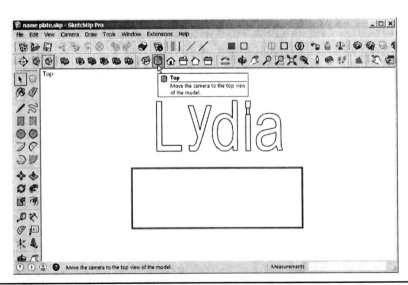

Figure 7-35   Lay the two groups flat, and generate a top view with Camera > Parallel Projection and the Top View icon.

Print Preview

Click Print Preview, and scroll to Adobe PDF in the first text screen to see what you're about to print (Figure 7-36). Here are the most relevant options:

- **Fit to Page.** Deselect this. It forces the print output to fit in a single page with no regard to scale. It also disables the Page Size fields and the Tiled Sheet Print Range options.

- **Use Model Extents.** Check this box to reduce the empty space around the model, thus reducing the number of tiled pages when printing large models. It doesn't affect the print's scale, just the overall size. This may help you to fit a scaled print onto a single page or reduce the number of pages the print requires. Figure 7-37 shows a .pdf made with Model Extents off, not a result we want.

You can print any size model. Small ones can be printed full scale and take up one page; larger ones need to be scaled down and will span multiple pages, as what 123D

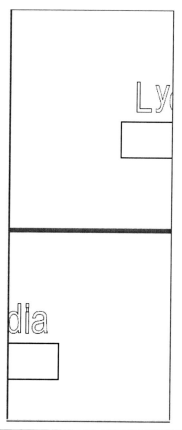

Figure 7-37 With Model Extents off, the model tiled across two pages. This is not a result we want.

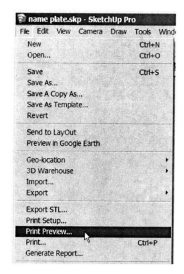

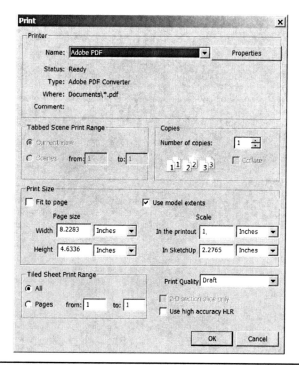

Figure 7-36 Click Print Preview, and scroll to Adobe PDF in the Name field to see what you're about to print.

Make generated. However, printing small models may result in a print that spans more pages than is necessary. Resolve this by manually resizing the SketchUp window to minimize the empty space around the model. Deselect Model Extents, click the Zoom Extents icon (Figure 7-38) to confirm the

Figure 7-38 The Zoom Extents icon fills the screen with the model and centers it.

model is centered, and then drag a corner of the SketchUp window to resize it, cropping as much empty space around the model as possible (Figure 7-39). You also can orient the page in portrait or landscape style to suit the dimensions of the model (discussed later in this chapter). Note how the Page Size and Tiled Sheet Print Range fields change as you toggle Use Model Extents on and off.

- **Scale.** This is the ratio of printed size to actual size. Look at the Scale fields (Figure 7-40). To make the print full size, type the

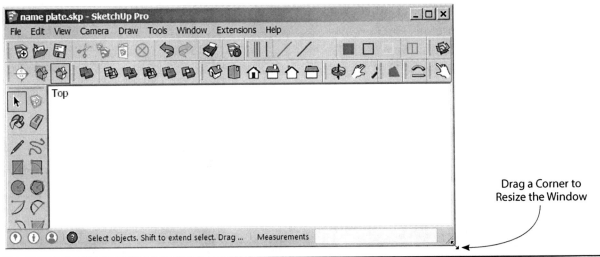

Figure 7-39 Drag a corner of the SketchUp window to crop unnecessary space.

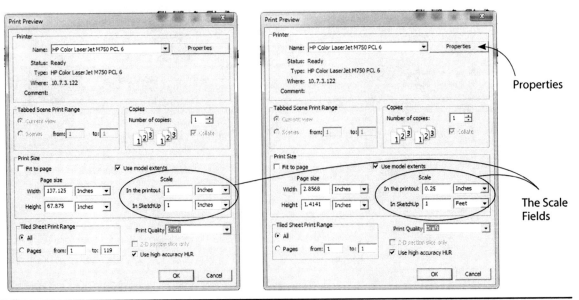

Figure 7-40 Type as shown in the text fields to print the model full scale (*left*) or ¼" = 1'0" (*right*).

following: *In the Printout*, type **1**, and select any unit from the adjacent field's drop-down menu. *In SketchUp*, type **1**, and select the same unit from the adjacent field's drop-down menu. To scale the print at 1/4" = 1'0", type the following: *In the Printout*, type **1/4** or **.25**, and choose Inches. *In SketchUp*, type **1** and choose Feet.

In both cases, make sure that Fit to Print is deselected, that Camera > Parallel Projection has been clicked, and that one of the Standard Views icons has been clicked. Otherwise, the scale fields will be gray.

■ **Page Size.** These fields display the overall size of the printing, not the size of the paper. When you type numbers into the Scale fields, SketchUp calculates the overall dimensions of the print and displays the resulting width and height in the Page Size fields. Don't type numbers into the Page Size fields when printing to scale. Look at the width and height dimensions displayed, and judge whether the printing will fit within the paper's printable area. Be aware that the Page Size width and height fields don't update automatically when you type numbers in the Scale fields. Toggle the Use Model Extents option to refresh the Page Size fields. Note the effect Use Model Extents has on the print's overall width and height.

■ **Tiled Sheet Print Range.** When the overall size of the printing is larger than the printable area of a single page, SketchUp spans, or tiles, the printing across multiple pages. You can print all the pages, a single page, or a range of pages. The default print range is All, and the From/To fields show the total number of pages. The Tiled Sheet

Print Range fields don't update automatically when you type numbers in the Scale fields. Refresh them by toggling the Use Model Extents option.

■ **Use High Accuracy HLR.** Check this box to create a vector file. Otherwise, SketchUp prints a raster file. Vector files only show colors, not textures. Textures and other images appear as shades. If you want the textures as part of the .pdf file, deselect this box.

Now click the Properties button to bring up another window (Figure 7-41). At the Adobe Pdf Settings tab, you can change the .pdf paper size, and at the Layout tab, you can orient the print as portrait or landscape (Figure 7-42). The Layout and Paper/Quality tabs both have advanced settings that you might want to explore for your particular project.

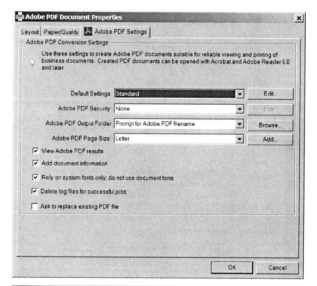

Figure 7-41 The Properties button brings up this window.

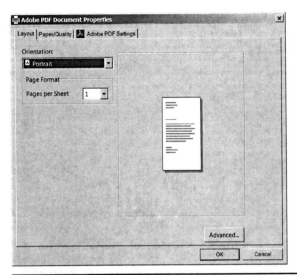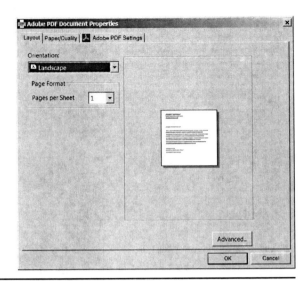

Figure 7-42   From the Layout tab, orient the print as portrait or landscape style.

Print the .pdf File

If you don't like the Print Preview image, click Close. Adjust the settings, and view it again. If you like it, click Print. The Print dialog box will open. Click OK. Figure 7-43 shows a .pdf file of the model's top view. It is a scaled orthographic drawing.

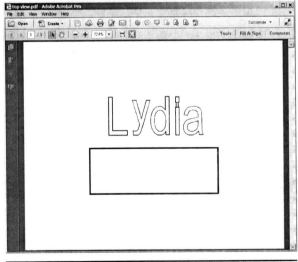

Figure 7-43   The .pdf file shows the model's top view.

Exporting a .dxf File

 At File > Export > 2D Graphic, SketchUp Pro has a .dxf option (Figure 7-44). This creates a vector file that can be imported into many CAM and Autodesk programs as well. As an aside, if you plan to give the file to a third party to fabricate, the .dxf format is often preferred because the machine operator knows it's a vector file instead of a possible raster file.

Make sure that Camera > Parallel Projection and one of the Standard Views icons are checked. Then click on the Options button. Most of these settings apply only if you want to export your SketchUp model to an Autodesk program for further development. At the end of this chapter is a link that explains them. Following are the ones relevant for CNC work:

- **Drawing Scale and Size.** This contains scaling options:
  - *Full Scale* (1:1) makes the drawing true-life size.
  - *In Drawing/In Model* scales the drawing. For example, to scale the drawing at 1/4" = 1", type the following: *In Model*, type **1"**. *In Drawing*, type **4'**.

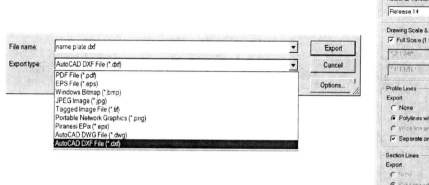

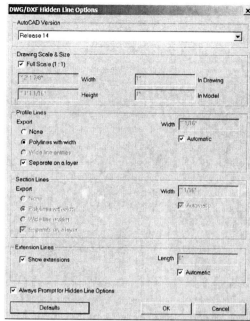

Figure 7-44 The model can be exported as a .dxf file. Click on the Options button for settings.

- **Width/Height.** Enter a custom page size for your file.
- **Profile Lines.** Check this to export any profile lines.
- **Section Lines.** This has options for exporting section lines.

Figure 7-45 shows the .dxf file open in AutoCAD. If you don't have an Autodesk program on your computer with which to see it, you can download a free .dwg viewer. A link is given at the end of this chapter.

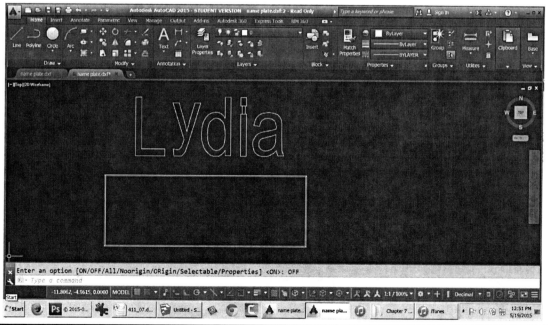

Figure 7-45 The exported .dxf file open in AutoCAD.

G-Code

After creating a .pdf or .dxf vector file, the next step is to import it into a CAM program that generates G-code. G-code, also called *post-processing*, tells the router the tool path, or how to move to make the part. It also tells the router how fast to move, what kind of stock it's cutting, the stock's thickness, and the finish type (such as rough or smooth). All CNC machines are built and configured differently, so their G-code is slightly different.

Generating G-code requires a working knowledge of speeds, feed rates, tool paths, cutter types, stepovers, the work-plane coordinate system, origin location, collision checks, tabs, 2 1/2D versus 3D, and properties of stock materials. This is beyond the scope of this book. There are busy online forums devoted to CNCing and CAM programs, and some are listed at the end of this chapter. However, we'll take a look at what a .pdf file looks like in one program and discuss issues that may arise.

Prepare the SketchUp-Exported .pdf Inside Vetric Cut2D

 Once you generate a .pdf or .dxf file, the next step is to import it into a CAM program that converts it to G-code. Let's import the name plate .pdf into a popular program called *Cut2D*. We're not going to go through this program's steps. Rather, I want to point out issues to watch for that will apply to any program you import it into.

First, import the file, and enter the sheet material's dimensions and location on the router bed (Figure 7-46).

A SketchUp-exported file almost always needs editing inside CAM software. Because of the nature of a 3D model and SketchUp, the imported file will almost never be machine-ready. For example, all shapes in a CNC file need to be closed polygons, but here we see that the base

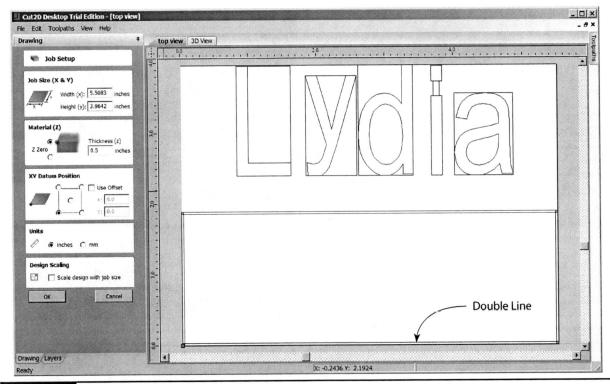

Figure 7-46 Import the .pdf file.

plate has a double line. In Chapter 3 we beveled it (Figure 7-47), and those double lines represent the bevel. But double lines make no sense to a router; the base must have just one line so that the router can tell which is outside and which is inside. So we need to delete the interior line, and can do so with Cut2D's editing tools (Figure 7-48).

Now look at the letters. Clicking on each one shows that they're not closed polygons; rather, they're made of multiple individual lines (Figure 7-49). This must be fixed, too. Additionally, SketchUp files often import with multiple lines stacked on top of each other. This will cause the router to repeat a cut in that location as many times as there are lines. So each line needs to be clicked on and deleted to ensure that there isn't

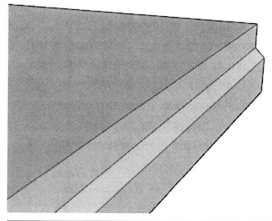

Figure 7-47 The base plate's beveled edge.

an identical line beneath it. Removing stacked lines can be done in the Cut2D software, but on a complex model it will take awhile to do. The *i* needed its top two pieces removed because

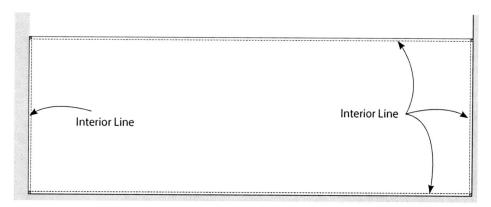

Interior Line

Interior Line

Figure 7-48 The interior edges must be deleted.

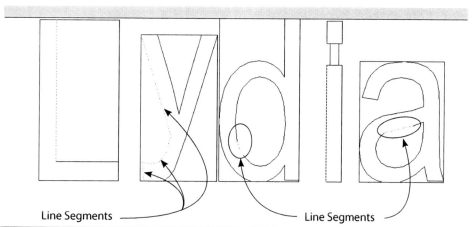

Line Segments

Line Segments

Figure 7-49 Each letter is composed of multiple lines. They must be edited to be closed polygons inside the CAM software.

Cut2D only has editing, not drawing, tools, and it would be difficult or impossible to turn all three into one closed polygon. Because of these kinds of issues, many Makers prefer to create their CNC designs in 2D drawing software instead of 3D software. Illustrator (Adobe), AutoCAD (Autodesk), Solidworks (Dassault), and Draw (Corel) are popular programs.

Once all work on the file is completed, G-code is generated. The file is saved and then transferred to the computer on the CNC machine, where it is read and cut. Figure 7-50 shows Mach 3 software, a popular G-code reader, and a close-up of the name plate file. The stock material is clamped down to the bed, and cutting begins. Figure 7-51 shows the name plate being cut out of a piece of foam. Note the tab holding the cut-out piece to the stock piece. This is one of the features you program into G-code. Figure 7-52 shows all the cut-out pieces, including the tabs that held the letters to the foam.

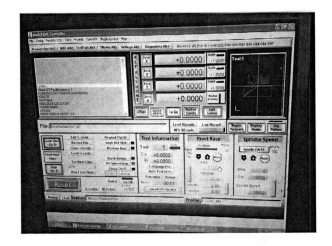

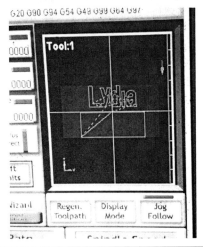

Figure 7-50 The Cut2D file opened in Mach 3.

Clamp

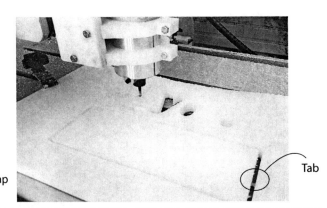

Tab

Figure 7-51 The CNC router reads the G-code and cuts the project.

Figure 7-52   The cut-out pieces.

The Right Tool for the Right Job

Figure 7-53 shows the CNC router cutting the project out of wood. However, it takes about four times as long to cut the name plate base on a CNC router than on a table saw (Figure 7-54). And while a decorative edge could be carved on the CNC router, a table router will do that job quicker, too (Figure 7-55). However, the letters are easiest and safest to cut out with the CNC router. Knowing which tools to use for a job is a large part of the job, too.

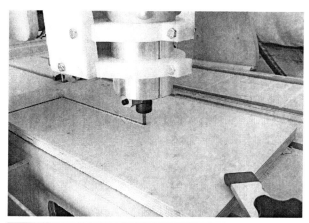

Figure 7-53   A CNC router cutting the base plate out of wood.

Figure 7-55   An ogee curve is cut on the name plate's edges with a table router. A vacuum is hooked up to the router to remove wood chips. (*Hammerspace, Kansas City, MO*)

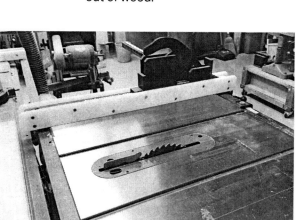

Figure 7-54   A table saw will do the job faster than a CNC router.

SketchUp LayOut

One more thing! If you want scaled, 2D, annotated construction documents of your model, SketchUp Pro's LayOut feature will do the job. It downloads as a separate program when you install SketchUp. Export the SketchUp model directly to LayOut via the LayOut icon (Figure 7-56) and create viewports, each of which can display and scale a different view of the model (Figure 7-57). You can also dimension and annotate the views (Figure 7-58). The document can be exported as a .pdf, .dwg., or dxf file. A video showing how to use LayOut is at https://youtu.be/izk5eWuubFs.

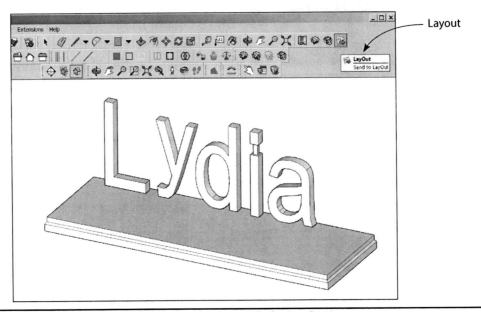

Figure 7-56 Import the SketchUp model directly into LayOut with the LayOut icon.

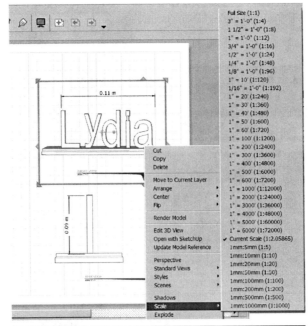

Figure 7-57 Create viewports, choose views, and scale the views.

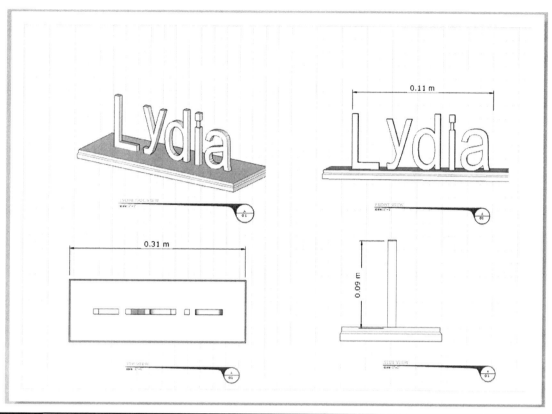

Figure 7-58 The final LayOut document.

Summary

In this chapter we learned how to turn a SketchUp model into a file for fabrication on a CNC router. We sliced a 3D .stl file into multiple 2D pattern pieces with Autodesk 123D Make. We generated scaled, orthographic .pdf and .dxf files with SketchUp Pro. We opened a SketchUp-exported .pdf file in Vetric Cut2D to work on the model further, then opened the G-code generated there in Mach 3 CAM software. We cut the pattern out in foam on a CNC router and looked at some manual tools that could be part of the workflow.

Resources

- **Video of the Terrain Model project:**
  https://www.youtube.com/watch?v=y22kYRnsuus

- **Video of the Nameplate under construction:**
  https://www.youtube.com/watch?v=XTxZsQFSzWs

- **Forum for CNC hobbyists and professionals:**
  http://cnczone.com

- **Forum for Vetric software users:**
  http://forum.vectric.com/

- **Autodesk CAM site:**
  http://cam.autodesk.com/

- **Service bureau for CNC projects:**
  http://ponoko.com

- **Interesting CNC projects:** http://blog.ponoko .com/2012/01/11/ten-excellent-examples-of -cnc-routing/

- **Find CNC-related extensions in the Extension Warehouse:** https://extensions.sketchup.com/

- **Explanation of .dxf export options:** http://help.sketchup.com/en/article/114293

- **Download a free .dwg viewer:** http://www.autodesk.com/products/dwg/ viewers#

Index

Page numbers in italics refer to figures.